CSET® Math CTC Skill Practice

Practice Test Questions for CSET® Mathematics Test

Copyright © 2016 by Complete Test Preparation Inc. ALL RIGHTS RESERVED.

No part of this book may be reproduced or transferred in any form or by any means, graphic, electronic, or mechanical, including photocopying, recording, web distribution, taping, or by any information storage retrieval system, without the written permission of the author.

Notice: Complete Test Preparation Inc. makes every reasonable effort to obtain from reliable sources accurate, complete, and timely information about the tests covered in this book. Nevertheless, changes can be made in the tests or the administration of the tests at any time and Complete Test Preparation Inc. makes no representation or warranty, either expressed or implied as to the accuracy, timeliness, or completeness of the information contained in this book. Complete Test Preparation Inc. make no representations or warranties of any kind, express or implied, about the completeness, accuracy, reliability, suitability or availability with respect to the information contained in this document for any purpose. Any reliance you place on such information is therefore strictly at your own risk.

The author(s) shall not be liable for any loss incurred as a consequence of the use and application, directly or indirectly, of any information presented in this work. Sold with the understanding, the author is not engaged in rendering professional services or advice. If advice or expert assistance is required, the services of a competent professional should be sought.

The company, product and service names used in this publication are for identification purposes only. All trademarks and registered trademarks are the property of their respective owners. Complete Test Preparation Inc. is not affiliated with any educational institution.

We strongly recommend that students check with exam providers for up-to-date information regarding test content.

CSET[®] is a registered trademark of National Evaluation Systems, Inc., who are not involved in the production of, and do not endorse this publication.

Published by
Complete Test Preparation Inc.
Victoria BC Canada

Visit us on the web at https://www.test-preparation.ca
Printed in the USA

ISBN-13: 9781772454826

Version 8 May 2025

About Complete Test Preparation Inc.

Why Us?
The Complete Test Preparation Team has been publishing high quality study materials since 2005, with a catalog of over 145 titles, in English, French and Chinese, as well as ESL curriculum for all levels.

To keep up with the industry changes we update everything all the time!

And the best part?
With every purchase, you're helping people all over the world improve themselves and their education. So thank you in advance for supporting this mission with us! Together, we are truly making a difference in the lives of those often forgotten by the system.

Charities that we support -
https://www.test-preparation.ca/charities-and-non-profits/

You have definitely come to the right place.
If you want to spend your valuable study time where it will help you the most - we've got you covered today and tomorrow.

Feedback

We welcome your feedback. Email us at feedback@test-preparation.ca with your comments and suggestions. We carefully review all suggestions and often incorporate reader suggestions into upcoming versions. As a Print on Demand Publisher, we update our products frequently.

 **Find us on Facebook**
www.facebook.com/CompleteTestPreparation

The Environment and Sustainability

Environmental consciousness is important for the continued growth of our company. In addition to eco-balancing each title, as a print on demand publisher, we only print units as orders come in, which greatly reduces excess printing and waste. This revolutionary printing technology also eliminates carbon emissions from trucks hauling boxes of books everywhere to warehouses. We also maintain a commitment to recycling any waste materials that may result from the printing process. We continue to review our manufacturing practices on an ongoing basis to ensure we are doing our part to protect and improve the environment.

Contents

6 **Getting Started**
   The CSET® Mathematics Study Plan — 7

12 **Practice Test Questions Set 1 (Easy)**
   Skills and Competencies — 42
   Answer Key — 47

79 **Practice Test Questions Set 2 (Difficult)**
   Skills and Competencies — 108
   Answer Key — 112

153 **Conclusion**

Getting Started

CONGRATULATIONS! By deciding to take the CSET® Mathematics Test, you have taken the first step toward a great future! Of course, there is no point in taking this important examination unless you intend to do your best to earn the highest grade you possibly can. That means getting yourself organized and discovering the best approaches, methods and strategies to master the material. Yes, that will require real effort and dedication on your part, however, if you are willing to focus your energy and devote the study time necessary, before you know it you will be on you will be passing your exam with a great mark!

We know that taking on a new endeavour can be scary, and it is easy to feel unsure of where to begin. That's where we come in.

About the Exam

The CSET® Mathematics Test is composed of three sub-tests:

- Subtest 1 - Numbers and Quantity, Algebra
- Subtest II - Geometry, Probability and Statistics
- Subtest III - Calculus

While we seek to make our guide as comprehensive as possible, note that like all exams, the CSET® Mathematics Test might be adjusted at some future point. New material might be added, or content that is no longer relevant or applicable might be removed. It is always a good idea to give the materials you receive when you register to take the CSET® Mathematics a careful review.

The CSET® Mathematics Study Plan

Now that you have made the decision to take the CSET® Mathematics, it is time to get started. Before you do another thing, you will need to figure out a plan of attack. The very best study tip is to start early! The longer the time period you devote to regular study practice, the more likely you will be to retain the material and be able to access it quickly. If you thought that 1x20 is the same as 2x10, guess what? It really is not, when it comes to study time. Reviewing material for just an hour per day over the course of 20 days is far better than studying for two hours a day for only 10 days. The more often you revisit a particular piece of information, the better you will know it. Not only will your grasp and understanding be better, but your ability to reach into your brain and quickly and efficiently pull out the tidbit you need, will be greatly enhanced as well.

The great Chinese scholar and philosopher Confucius believed that true knowledge could be defined as knowing both what you know and what you do not know. The first step in preparing for the CSET® Mathematics is to assess your strengths and weaknesses.

Making a Study Schedule

To make your study time the most productive, you will need to develop a study plan. The purpose of the plan is to organize all the bits of pieces of information in such a way that you will not feel overwhelmed. Rome was not built in a day, and learning everything you will need to know to pass the CSET® Mathematics is going to take time, too. Arranging the material you need to learn into manageable chunks is the best way to go. Each study session should make you feel as though you have succeeded in accomplishing your goal, and your goal is simply to learn what you planned to learn during that particular session. Try to organize the content in such a way that each study session builds on previous ones. That way, you will retain the information, be better able to access it, and review the previous bits and pieces at the same time.

Self-assessment

The Best Study Tip! The very best study tip is to start early! The longer you study regularly, the more you will retain and 'learn' the material. Studying for 1 hour per day for 20 days is far better than studying for 2 hours for 10 days.

What don't you know?

The first step is to assess your strengths and weaknesses. You may already have an idea of where your weaknesses are, or you can take our Self-assessment modules for each of the areas, Reading Comprehension, Arithmetic, Essay Writing, Algebra and College Level Math.

Exam Component	Rate 1 to 5
Number and Quantity	
Algebra	
Calculus	
Geometry	
Probability and Statistics	

Making a Study Schedule

The key to a study plan is to divide the material you need to learn into manageable size and learn it, while at the same time reviewing the material that you already know.

Using the table above, any scores of three or below, you need to spend time learning, going over and practicing this subject area. A score of four means you need to review the material, but you don't have to spend time re-learning. A score of five and you are OK with just an occasional review before the exam.

A score of zero or one means you really do need to work on this and you should allocate the most time and give it the highest priority. Some students prefer a 5-day plan and others a 10-day plan. It also depends on how much time until the exam.

Here is an example of a 5-day plan based on an example from the table above:

Number and Quantity: 1   Study 1 hour everyday – review on last day
Algebra: 3   Study 1 hour for 2 days then ½ hour and then review
Calculus: 4   Review every second day
Geometry: 2 Study 1 hour on the first day – then ½ hour everyday
Probability and Statistics: 5   Review for ½ hour every other day

Using this example, Geometry and Probability and Statistics is good, and only need occasional review. Calculus is good and needs 'some' review. Algebra needs a bit of work and Number and Quantity is very weak and needs the most time. Based on this, here is a sample study plan:

Day	Subject	Time
Monday		
Study	Number and Quantity	1 hour
Study	Geometry	1 hour
	½ hour break	
Study	Algebra	1 hour
Review	Calculus	½ hour
Tuesday		
Study	Number and Quantity	1 hour
Study	Geometry	½ hour
	½ hour break	
Study	Algebra	½ hour
Review	Calculus	½ hour
Wednesday		
Study	Number and Quantity	1 hour
Study	Geometry	½ hour
	½ hour break	
Study	Algebra	½ hour
Thursday		
Study	Number and Quantity	½ hour
Study	Geometry	½ hour
Review	Algebra	½ hour
	½ hour break	
Review	Calculus	½ hour
Friday		
Review	Number and Quantity	½ hour
Review	Geometry	½ hour
Review	Algebra	½ hour
	½ hour break	
Review	Calculus	½ hour
Review	Geometry	½ hour

Using this example, adapt the study plan to your own schedule. This schedule assumes 2 ½ - 3 hours available to study everyday for a 5 day period.

First, write out what you need to study and how much. Next figure out how many days before the test. Note, do NOT study on the last day before the test. On the last day before the test, you won't learn anything and will probably only confuse yourself.

Make a table with the days before the test and the number of hours you have available to study each day. We suggest working with 1 hour and ½ hour time slots.

Start filling in the blanks, with the subjects you need to study the most getting the most time and the most regular time slots (i.e. everyday) and the subjects that you know getting the least time (e.g. ½ hour every other day, or every 3rd day).

Tips for making a schedule

Once you make a schedule, stick with it! Make your study sessions reasonable. If you make a study schedule and don't stick with it, you set yourself up for failure. Instead, schedule study sessions that are a bit shorter and set yourself up for success! Make sure your study sessions are do-able. Studying is hard work, but after you pass, you can party and take a break!

Schedule breaks. Breaks are just as important as study time. Work out a rotation of studying and breaks that works for you.

Build up study time. If you find it hard to sit still and study for 1 hour straight-through, build up to it. Start with 20 minutes, and then take a break. Once you get used to 20-minute study sessions, increase the time to 30 minutes. Gradually work you way up to 1 hour.

40 minutes to 1 hour is optimal. Studying for longer than this is tiring and not productive. Studying for shorter isn't long enough to be productive.

Skills and Competencies

Subtest 1
Number and Quantity

1. Perform operations with fractions

2. Perform operations with fractions

3. Simplify and approximate radicals

4. Solve equations with rational or radical expressions

5. Solve problems with ratio and velocity

6. Fractions, decimals and percent

7. Fractions, decimals and percent

8. Solve equations using exponents

9. Solve equations using exponents

10. Solve problems using factoring, greatest common factor, least common multiple

Algebra

11. Perform operations with polynomials

12. Perform operations with polynomials

13. Solve quadratic equations

Skills and Competencies

14. Solve quadratic equations

15. Solve problems using graphs of quadratics

16. Solve systems of linear equations in 2 variables

17. Solve linear equations in 1 variable

18. Solve linear equations in 1 variable

19. Simplify expressions with radicals and rational exponents using properties of exponents

20. Identify graphs of linear equations or inequality with 2 variables on coordinate plane

21. Rewrite and manipulate rational expressions

22. Evaluate whether a particular mathematical model (e.g., graph, equation, table) describes a given set of conditions

23. Evaluate whether a particular mathematical model (e.g., graph, equation, table) describes a given set of conditions.

24. Rewrite and manipulate rational expressions

25. Develop a model (e.g., graph, equation, table) of a given set of conditions

26. Calculate the domain of a function

27. Calculate arithmetic and geometric sequences

28. Determine the domain and range from a given graph of a function

29. Understand and use sequences and recursive functions

30. Solve logarithmic and exponential functions

31. Model periodic phenomena with trigonometric functions

32. Perform operations with functions

33. Perform operations with functions

34. Understand and calculate inverse functions

35. Understand and calculate inverse functions

Subtest II Geometry

1. Solve problems with Pythagorean geometry

2. Determine congruence

3. Reflect geometric shapes

4. Determine congruence

5. Reflect geometric shapes

6. Central angles, tangents, arcs, and sectors

7. Calculate the volume of a cone

8. Determine the equation of a conic section to model real-world situations

9. Calculate the volume of a cylinder

10. Calculate the diameter and volume of a circle

11. Solve problems with Pythagorean geometry

12. Calculate the volume of a cylinder

13. Calculate surface area

14. Solve problems with paralell lines

15. Calculate the area of a circle

Probability and Statistics

16. Calculate probability with dependent values

17. Calculate the mean of a set of data

18. Calculate simple probability

19. Calculate the median of a set of numbers

20. Calculate mode of a set of data

21. Calculate independence and conditional probability

22. 1 and 2 variable data in different formats

23. Correlation and causation

24. Make inferences about a population from a single random sample

25. Use probability/permutations and combinations to compute probabilities of compound events

Subtest III Calculus

1. Calculate definite integrals

rst derivative of a function

tives and definite integrals as limits (differ-

4. Understand first derivatives of a function

5. Calculate power series

6. Use derivatives - to solve rates of change

7. Use first order differential equations to solve separation of variables and initial value problems.

8. Calculate the derivative of a function as a limit

9. Use the fundamental theorem of calculus to solve problems

10. Find local minimum and maximums of functions

11. Interpret derivatives and definite integrals as limits (difference quotients, slope, Riemann sums area)

12. Determine limits using theorems concerning sums, products and quotients of functions

13. Determine the slope or equation of a tangent line at a point on a curve

14. Use the Determine the first derivative of a function in various representations to determine increasing and decreasing intervals or extrema

15. Solve distance, area, and volume problems using integration

Practice Test Questions Set 1 (Easy)

The questions below are not the same as you will find on the CSET® Mathematics - that would be too easy! And nobody knows what the questions will be and they change all the time. Below are general questions that cover the same subject areas as the CSET® Mathematics test. So, while the format and exact wording of the questions may differ slightly, and change from year to year, if you can answer the questions below, you will have no problem with the CSET® Mathematics test.

For the best results, take these practice test questions as if it were the real exam. Set aside time when you will not be disturbed, and a location that is quiet and free of distractions. Read the instructions carefully, read each question carefully, and answer to the best of your ability.
Use the bubble answer sheets provided. When you have completed the practice questions, check your answer against the Answer Key and read the explanation provided.

Do not attempt more than one set of practice test questions in one day. After completing the first practice test, wait two or three days before attempting the second set of questions.

Subtest I Answer Sheet

1. A B C D   18. A B C D
2. A B C D   19. A B C D
3. A B C D   20. A B C D
4. A B C D   21. A B C D
5. A B C D   22. A B C D
6. A B C D   23. A B C D
7. A B C D   24. A B C D
8. A B C D   25. A B C D
9. A B C D
10. A B C D
11. A B C D
12. A B C D
13. A B C D
14. A B C D
15. A B C D
16. A B C D
17. A B C D

Subtest II Answer Sheet

1. Ⓐ Ⓑ Ⓒ Ⓓ
2. Ⓐ Ⓑ Ⓒ Ⓓ
3. Ⓐ Ⓑ Ⓒ Ⓓ
4. Ⓐ Ⓑ Ⓒ Ⓓ
5. Ⓐ Ⓑ Ⓒ Ⓓ
6. Ⓐ Ⓑ Ⓒ Ⓓ
7. Ⓐ Ⓑ Ⓒ Ⓓ
8. Ⓐ Ⓑ Ⓒ Ⓓ
9. Ⓐ Ⓑ Ⓒ Ⓓ
10. Ⓐ Ⓑ Ⓒ Ⓓ
11. Ⓐ Ⓑ Ⓒ Ⓓ
12. Ⓐ Ⓑ Ⓒ Ⓓ
13. Ⓐ Ⓑ Ⓒ Ⓓ
14. Ⓐ Ⓑ Ⓒ Ⓓ
15. Ⓐ Ⓑ Ⓒ Ⓓ
16. Ⓐ Ⓑ Ⓒ Ⓓ
17. Ⓐ Ⓑ Ⓒ Ⓓ

Subtest III Answer Sheet

1. Ⓐ Ⓑ Ⓒ Ⓓ
2. Ⓐ Ⓑ Ⓒ Ⓓ
3. Ⓐ Ⓑ Ⓒ Ⓓ
4. Ⓐ Ⓑ Ⓒ Ⓓ
5. Ⓐ Ⓑ Ⓒ Ⓓ
6. Ⓐ Ⓑ Ⓒ Ⓓ
7. Ⓐ Ⓑ Ⓒ Ⓓ
8. Ⓐ Ⓑ Ⓒ Ⓓ
9. Ⓐ Ⓑ Ⓒ Ⓓ
10. Ⓐ Ⓑ Ⓒ Ⓓ
11. Ⓐ Ⓑ Ⓒ Ⓓ
12. Ⓐ Ⓑ Ⓒ Ⓓ
13. Ⓐ Ⓑ Ⓒ Ⓓ
14. Ⓐ Ⓑ Ⓒ Ⓓ
15. Ⓐ Ⓑ Ⓒ Ⓓ
16. Ⓐ Ⓑ Ⓒ Ⓓ
17. Ⓐ Ⓑ Ⓒ Ⓓ
18. Ⓐ Ⓑ Ⓒ Ⓓ
19. Ⓐ Ⓑ Ⓒ Ⓓ
20. Ⓐ Ⓑ Ⓒ Ⓓ
21. Ⓐ Ⓑ Ⓒ Ⓓ
22. Ⓐ Ⓑ Ⓒ Ⓓ
23. Ⓐ Ⓑ Ⓒ Ⓓ
24. Ⓐ Ⓑ Ⓒ Ⓓ
25. Ⓐ Ⓑ Ⓒ Ⓓ
26. Ⓐ Ⓑ Ⓒ Ⓓ
27. Ⓐ Ⓑ Ⓒ Ⓓ
28. Ⓐ Ⓑ Ⓒ Ⓓ
29. Ⓐ Ⓑ Ⓒ Ⓓ
30. Ⓐ Ⓑ Ⓒ Ⓓ
31. Ⓐ Ⓑ Ⓒ Ⓓ
32. Ⓐ Ⓑ Ⓒ Ⓓ
33. Ⓐ Ⓑ Ⓒ Ⓓ
34. Ⓐ Ⓑ Ⓒ Ⓓ
35. Ⓐ Ⓑ Ⓒ Ⓓ

Part I - Number and Quantity

1. Solve 2/3 + 5/12

    a. 9/17
    b. 3/11
    c. 7/12
    d. 1 1/12

2. 15/16 x 8/9

    a. 5/6
    b. 16/37
    c. 2/11
    d. 5/7

3. If $x = \sqrt{7} - 1$ and $y = \sqrt{7} + 1$, find the value of (x + y) / (x - y).

    a. $-\sqrt{7}$
    b. -2
    c. 2
    d. $\sqrt{7}$

4. We are given that A = $(\sqrt{3} - 1) / (\sqrt{5} + 1)$ and B = $(\sqrt{5} - 1) / (\sqrt{3} + 1)$. What is the value of A in terms of B?

    a. B/2
    b. 3B/2
    c. 2B
    d. 3B

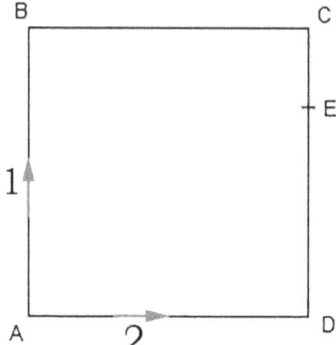

5. Two bicycles start moving from point A; one in direction 1, the other in direction 2. They first meet at point E. We know that 4|CE| = |CD| and ABCD is square shaped. Find the ratio of the velocities of the bicycles V1/V2.

    a. 1/7
    b. 3/7
    c. 5/7
    d. 9/7

6. 15 is what percent of 200?

    a. 7.5%
    b. 15%
    c. 20%
    d. 17.50%

7. Convert 75% to a fraction.

    a. 2/100
    b. 85/100
    c. 3/4
    d. 4/7

8. If x = 2 and y = 5, solve $xy^3 - x^3$

　　a. 240
　　b. 258
　　c. 248
　　d. 242

9. $X^3 * X^2 =$

　　a. 5^x
　　b. x^{-5}
　　c. x^{-1}
　　d. x^5

10. Find the number of positive factors of 360.

　　a. 24
　　b. 25
　　c. 32
　　d. 36

Algebra

11. Turn the following expression into a simple polynomial: (a + b)(x + y) + (a - b)(x - y) - (ax + by)

　　a. ax + by
　　b. ax - by
　　c. $ax^2 + by^2$
　　d. $ax^2 - by^2$

12. Given polynomials $A = 4x^5 - 2x^2 + 3x - 2$ and $B = -3x^4 - 5x^2 - 4x + 5$, find $A + B$.

   a. $x^5 - 3x^2 - x - 3$
   b. $4x^5 - 3x^4 + 7x^2 + x + 3$
   c. $4x^5 - 3x^4 - 7x^2 - x + 3$
   d. $4x^5 - 3x^4 - 7x^2 - x - 7$

13. Using the factoring method, solve the quadratic equation: $x^2 + 4x + 4 = 0$

   a. 0 and 1
   b. 1 and 2
   c. 2
   d. -2

14. Using the quadratic formula, solve the quadratic equation: $x - 31/x = 0$

   a. $-\sqrt{13}$ and $\sqrt{13}$
   b. $-\sqrt{31}$ and $\sqrt{31}$
   c. $-\sqrt{31}$ and $2\sqrt{31}$
   d. $-\sqrt{3}$ and $\sqrt{3}$

15. Find the x-intercepts of the quadratic function $f(x) = (x - 5)^2 - 9$.

   a. {2, 4}
   b. {2, 8}
   c. {4, 8}
   d. {1, 2}

16. Solve the system: $4x - y = 5$; $x + 2y = 8$

    a. (3, 2)
    b. (3, 3)
    c. (2, 3)
    d. (2, 2)

17. Solve the linear equation: $-x - 7 = -3x - 9$

    a. -1
    b. 0
    c. 1
    d. 2

18. Solve the linear equation:
$3(x + 2) - 2(1 - x) = 4x + 5$

    a. -1
    b. 0
    c. 1
    d. 2

19. If $2^{x-1} = 3$, find the value of 8^x.

    a. 16
    b. 36
    c. 186
    d, 216

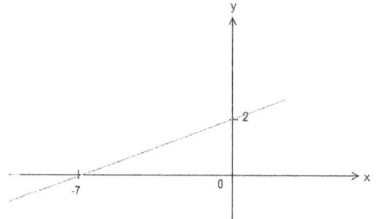

20. Which of the following is the equation of the function graphed above?

    a. $2x - 7y - 14 = 0$
    b. $2y - 7x + 14 = 0$
    c. $2x - 7y + 14 = 0$
    d. $2y - 7x - 14 = 0$

21. Simplify the expression $((x - y)^7 + x^2 + y^2 + (y - x)^7 + 2xy) / (y^2 - x^2)$.

    a. $(x + y) / (x - y)$
    b. $(x + y) / (y - x)$
    c. $(x - y) / (x + y)$
    d. $(y - x) / (x + y)$

22. Find the equation of a circle with center coordinates (2, 5) and radius 3.

    a. $2x^2 + 5y^2 - 4x - 10y + 20 = 0$
    b. $5x^2 + 2y^2 - 4x - 10y + 20 = 0$
    c. $x^2 + y^2 - 4x - 10y + 9 = 0$
    d. $x^2 + y^2 - 4x - 10y + 20 = 0$

23. Which of the following does not determine a function?

a. f = {(0, 2), (1, 5), (5, 5), (3, 0)}
b. y = 4x²

c.

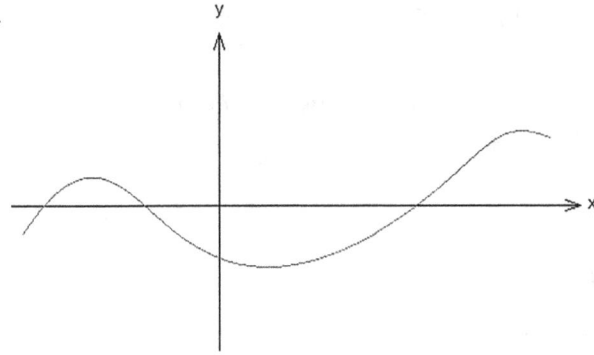

d.

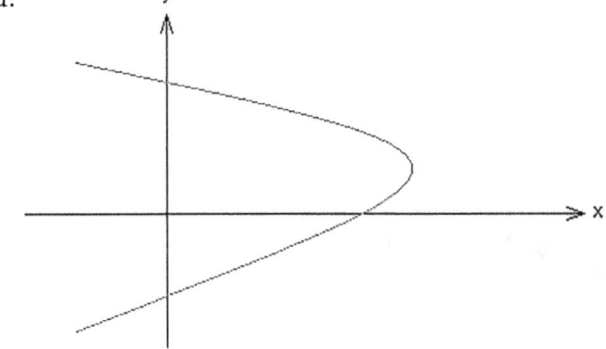

24. Simplify the expression
$((x - y)^7 + x^2 + y^2 + (y - x)^7 + 2xy) / (y^2 - x^2)$.

a. $(x + y) / (x - y)$
b. $(x + y) / (y - x)$
c. $(x - y) / (x + y)$
d. $(y - x) / (x + y)$

25. According to the table below, develop an equation for y in terms of x.

x	y
2	6
4	24
6	54
8	96

a. $y = 3x$
b. $y = 9x$
c. $y = 2x^2$
d. $y = 3x^2 / 2$

26. Find the domain of the function $f(x) = \sqrt{(x + 7)} / (x - 3)$.

a. $[3, 7) \cup (7, +\infty)$
b. $(-\infty, 3) \cup (3, +\infty)$
c. $[-7, 3) \cup (3, +\infty)$
d. $[-7, 3) \cup (3, +\infty]$

27. Find the sum of the first 10 terms of the sequence 3, 6, 12, ...

    a. 189
    b. 765
    c. 3069
    d. 6141

28. Find the range of the function given below:

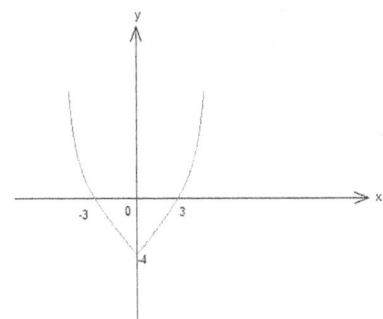

    a. [-3, 3]
    b. (-∞, +∞)
    c. (-∞, -3) ∪ (3, +∞)
    d. [-4, +∞)

29. The initial value for function f is given by f(1) = 3. The general formula of this function is f(x) = x * f(x - 1).

What is the value of f(20)?

    a. 3*20!
    b. 20^3
    c. 20*21
    d. 600

30. What is the result of $(\log_x y / \log_{xz} y^3) - \log_x \sqrt[3]{z}$?

   a. 1/3
   b. 1/27
   c. 3
   d. 9

31. A ball 32 cm above the floor is attached to the end of a spring attached to the ceiling. Initially, we pull the ball 6 cm down and when we let it move, it performs one up and down motion in 4 seconds. Modeling this harmonic movement using trigonometric functions (Assume that there is no air friction), find the distance between the ball and the ceiling at t = 9.5 seconds. Round your answer to the nearest hundredths.

   a. 28.12 cm
   b. 30.28 cm
   c. 32.36 cm
   d. 36.24 cm

32. Find g∘f if f(x) = 2x + 5 and g(x) = 5x + 2.

   a. 5x + 5
   b. 10x + 27
   c. 10x + 2
   d. 25x + 25

33. If f(x) = 1 + x², find f∘f .

   a. $1 + x^2 + x^4$
   b. $2 + x^2 + x^4$
   c. $2 + x^2$
   d. $1 + x^4$

34. Find f⁻¹(1/2) if f(x) = 1 - x.

   a. 1
   b. 1/2
   c. 1/3
   d. 1/4

35. If f(x) = 5x and g(x) = 7 - 2x, find (f - g)⁻¹(0).

   a. 1
   b. 2
   c. 3
   d. 4

Subtest II - Geometry and Data

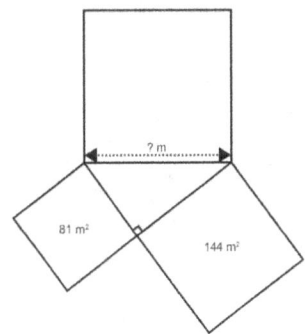

Note: figure not drawn to scale

1. What is the length of each side of the indicated square above? Assume the 3 shapes around the center triangle are square.

   a. 10
   b. 15
   c. 20

d. 5

2. For triangles ABC and A'B'C' we are given:

BC = B'C'
AC = A'C'

∠ A = ∠ A'

Are these 2 triangles congruent?

    a. Yes
    b. No
    c. Not enough information

3. Reflect the circle with the center in O with the given mirror line m.

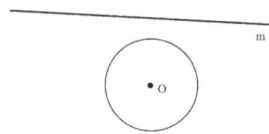

4. For triangles ABC and A'B'C' we have that:

AB = A'B'

∠ A = ∠ A'
∠ B = ∠ B'

Are these 2 triangles congruent?

    a. Yes
    b. No
    c. Not enough information

5. Reflect the rectangle ABCD with the given mirror line m in the space below.

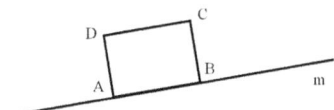

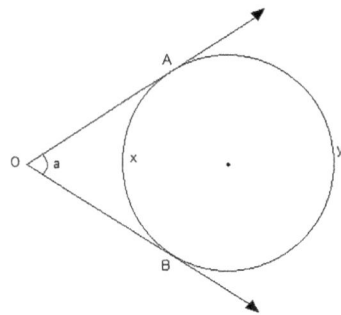

6. In the figure above, [OA and [OB are tangent to the circle and a = 60⁰. Find x.

    a. 120⁰
    b. 180⁰
    c. 220⁰
    d. 240⁰

7. Find the volume of a cone with radius 5 cm and height 12 cm in m³.

    a. $\pi * 10^{-6}$ m³
    b. $5\pi * 10^{-5}$ m³
    c. $\pi * 10^{-4}$ m³
    d. $12\pi * 10^{-3}$ m³

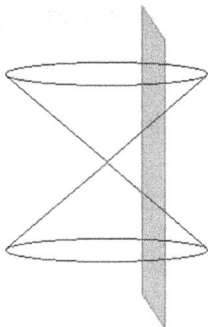

8. If we locate two cones vertex to vertex and cut with a razor blade as shown; which geometric shape do we obtain on the cross-section?

    a. Two circles
    b. Parabola
    c. Hyperbola
    d. Two ellipses

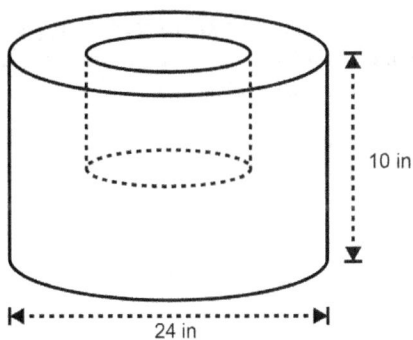

Note: Figure not drawn to scale

9. What is the volume of the above solid made by a hollow cylinder that is half the size (in all dimensions) of the larger cylinder?

   a. $1440 \pi \text{ in}^3$
   b. $1260 \pi \text{ in}^3$
   c. $1040 \pi \text{ in}^3$
   d. $960 \pi \text{ in}^3$

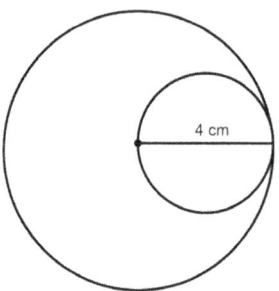

Note: Figure not drawn to scale

10. Assuming the diameter of the small circle is equal to the radius of the large circle, what is (area of large circle) - (area of small circle) in the figure above?

   a. $8 \pi \text{ cm}^2$
   b. $10 \pi \text{ cm}^2$
   c. $12 \pi \text{ cm}^2$
   d. $16 \pi \text{ cm}^2$

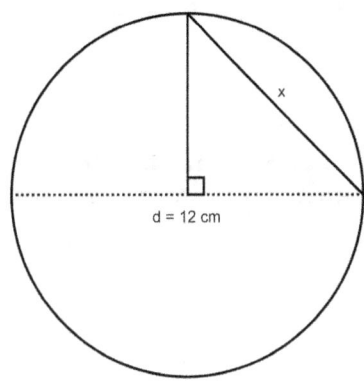

Note: figure not drawn to scale

11. Calculate the length of side x.

    a. 6.46
    b. 8.48
    c. 3.6
    d. 6.4

12. How much water can be stored in a cylindrical container 5 meters in diameter and 12 meters high?

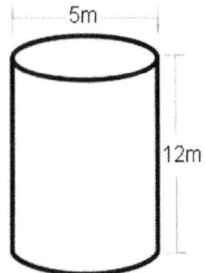

a. 235.65 m³
b. 223.65 m³
c. 240.65 m³
d. 252.65 m³

13. 3 boys are asked to clean a surface that is 4 ft². If the surface is divided equally among the boys, how much will each clean?

a. 1 ft 6 inches²
b. 14 inches²
c. 1 ft 2 inches²
d. 1 ft² 48 inches²

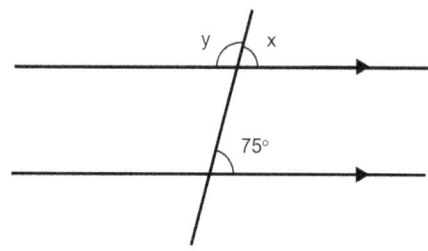

14. What is the value of the angle y?

a. 25°
b. 15°
c. 30°
d. 105°

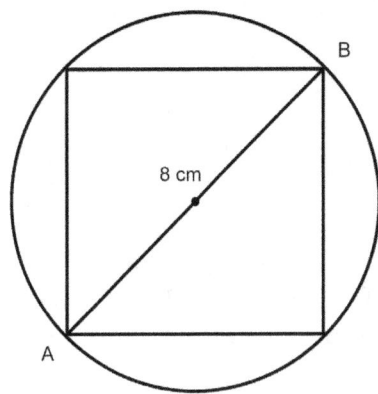

Note: figure not drawn to scale

15. What is area of the circle above?

a. 4π cm²
b. 12π cm²
c. 10π cm²
d. 16π cm²

Probability and Statistics

16. There are 3 blue, 1 white and 4 red identical balls inside a bag. If it is aimed to take two balls out of the bag consecutively, what is the probability to have 1 blue and 1 white ball?

a. 3/28
b. 1/12
c. 1/7
d. 3/7

17. Find the mean of these set of numbers:
100, 1050, 320, 600 and 150

   a. 333
   b. 444
   c. 440
   d. 320

18. A boy has 4 red, 5 green and 2 yellow balls. He chooses two balls randomly for play. What is the probability that one is red and other is green?

   a. 2/11
   b. 19/22
   c. 20/121
   d. 9/11

19. Find the median of the set of numbers:
1, 2, 3, 4, 5, 6, 7, 8, 9 and 10

   a. 55
   b. 10
   c. 1
   d. 5.5

20. The following represents the age distribution of students in an elementary class. Find the mode of the values: 7, 9, 10, 13, 11, 7, 9, 19, 12, 11, 9, 7, 9, 10, 11

   a. 7
   b. 9
   c. 10
   d. 11

21. There are 5 blue, 5 green and 5 red books on a shelf. Two books are selected randomly. What is the probability of choosing two books of different colors?

    a. 1/3
    b. 2/5
    c. 4/7
    d. 5/7

22. Which of the following can be used to represent 2 variable data?

    a. Pie graph
    b. Scatter plot
    c. Bar graph
    d. Histogram

23. Which of the following contain both correlation and causation?

    a. The more exhibitions on art are organized in a country, the more traffic accidents occur.

    b. The number of babies born decreases by the increasing number of books written on adventure.

    c. The tougher weather conditions increased heater sales.

    d. Shorter people are less likely to be killers.

24. A company, surveys a random sample of 120 employees about the number of days next month they prefer to eat out, instead of at the company cafeteria. 80 expected to eat out 5 days next month. There are 450 employees in the company. Based on the data, find the most reasonable estimate for the number of employees who expect to eat out 5 days next month.

    a. 240

    b. 300

    c. 360

    d. 390

25. How many different ways can a reader choose 3 books out of 4, ignoring the order of selection?

    a. 3

    b. 4

    c. 9

    d. 12

Subtest III - Calculus

1. Calculate the definite integral $\int 2x^{-3} dx$.

    a. -5/36

    b. -5/72

    c. 1/18

    d. 2

2. Given $G(x) = \int_{x}^{\pi/4} \cos^2 3x \, dx$, find $G'(\pi/6)$.

   a. 0
   b. 1
   c. 1/2
   d. √3 / 2

3. Given that $\int_{-1}^{3} f(x)dx = -15$ and $\int_{3}^{-1} g(x)dx = 8$, find the value of $\int_{-1}^{3} (4g(x) - 5f(x))dx$.

   a. -107
   b. -43
   c. 43
   d. 107

4. Which of the following statements is correct?

   a. If the first derivative of the function is positive, then the function is decreasing.

   b. If the first derivative of the function is negative, then the function is increasing.

   c. If the second derivative of the function is positive, then the function is concave down.

   d. If the second derivative of the function is negative, then the function is concave down.

5. Using the first 5 terms of the Taylor Series 1 / (1 - x), find the approximate value of 5/4. Round your answer to the nearest hundredths.

   a. 1.20
   b. 1.23
   c. 1.24
   d. 1.25

6. A cylindrical tank is filled with water with a rate of 8 cm³/min. Find the rate at which the height of the water is increasing. The radius of the tank is 2 cm.

   a. $2/\pi$
   b. 3
   c. π
   d. 2π

7. Using the method of separation of variables, find y in terms of x in $dy/dx = x^3/y^2$.

   a. $y = x^{3/2} + C$
   b. $y = \sqrt{(x^{2/3} + C)}$
   c. $y = \sqrt[3]{(3x^4/4 + 3C)}$
   d. $y = \sqrt[4]{(x^3/4 - 3C)}$

8. Using the limit definition, compute the derivative of $f(x) = 3x^2 - 2x + 7$.

   a. $6x^2 + 7$
   b. $6x - 2$
   c. $-2x + 7$
   d. $x^3 + 7x$

9. Given $F(x) = \int_{4}^{x} x^5 dx$, find $F'(3)$.

   a. 243
   b. 475
   c. 576
   d. 810

10. Find the local minimum of the function $6x^3 - 36x + 23$.

    a. (-2, -36)
    b. (2, 36)
    c. (-√2, -36√2)
    d. (√2, 36√2)

11. Find the difference between the integral of $f(x) = 2x^2$ and the area under this graph of this function using the Riemann sum with $\Delta x = 1$, within the interval [-3, 3].

    a. 12
    b. 15
    c. 18
    d. 20

12. Find the limit of the function $(\sin 3x * (2x^2 + 3x - 5)) / (6x^2 + 15x)$ while $x \to 0$.

    a. -1
    b. 1
    c. 3
    d. 15

13. Find the equation of the line tangent to the curve y = 3x³ - 8 at x = 1.

    a. y = 9x - 14
    b. y = -9x + 14
    c. y = 9x + 16
    d. y = 18x + 16

14. Find the local maximum of the function f(x) = x³ - 27x.

    a. 1
    b. 3
    c. 18
    d. 27

15. If the graph of f(x) = 3 + sin²x is rotated π/2 degrees around the x-axis, what will the volume of the forming solid be within [0, 2π]?

    a. $7\pi^2/12$
    b. $19\pi^2/4$
    c. $9\pi^2/2$
    d. $19\pi^2$

Answer Key

Subtest I
Number and Quantity

1. D
A common denominator is needed, number which both 3 and 12 will divide into. So, 8 + 5/12 = 13/12 = 1 1/12.

2. C
Since the denominators are the same, we can just subtract the numerators, so 13 - 7/15 = 6/15 = 2/5

3. A
First, insert the values of x and y into the expression given:
$(x + y) / (x - y) = (\sqrt{7} - 1 + \sqrt{7} + 1) / (\sqrt{7} - 1 - (\sqrt{7} + 1))$
$= (2\sqrt{7}) / (\sqrt{7} - 1 - \sqrt{7} - 1) = (2\sqrt{7}) / (-2) = -\sqrt{7}$

4. A
Notice the denominator of A and numerator of B are conjugates. If we equate the denominators of A and B with the same number; it is easier to write A in terms of B.

$A = (\sqrt{3} - 1) / (\sqrt{5} + 1)_{(\sqrt{3} + 1)}$

$= ((\sqrt{3} - 1) * (\sqrt{3} + 1)) / ((\sqrt{5} + 1) * (\sqrt{3} + 1)) = (3 - 1) / ((\sqrt{5} + 1) * (\sqrt{3} + 1))$

$= 2 / ((\sqrt{5} + 1) * (\sqrt{3} + 1))$

$B = (\sqrt{5} - 1) / (\sqrt{3} + 1)_{(\sqrt{5} + 1)}$

$= ((\sqrt{5} - 1) * (\sqrt{5} + 1)) / ((\sqrt{5} + 1) * (\sqrt{3} + 1))$

$= (5 - 1) / ((\sqrt{5} + 1) * (\sqrt{3} + 1))$

$= 4 / ((\sqrt{5} + 1) * (\sqrt{3} + 1))$

There is no need to expand the denominators of the new forms of A and B since they are the same. Comparing the numerators is sufficient. Notice that B is 2 times A. So A = B/2.

5. D
Bicycle 1 follows the path passing through points A - B - C - E. Bicycle 2 follows A - D - E. We are given that 4|CE| = |CD| and the path is square-shaped, so the length of one side is 4a. Then:

   1) Bicycle 1 goes 4a + 4a + a = 9a

   2) Bicycle 2 goes 4a + 3a = 7a distance. Since they meet at E; the ratio of their velocities is equal to the ratio of distances they have taken:

   3) V_1/V_2 = 9a/7a = 9/7

6. A
15/200 = X/100
200X = (15 * 100)
1500/200   Cancel zeros in the numerator and denominator
15/2 = 7.5%.

Notice that the questions asks, What 15 is what percent of 200? The question does *not* ask, what is 15% of 200! The answers are very different.

7. C
75% = 75/100 = 3/4

8. D
$2(5)^3 - (2)^3$ = 2(125) − 8 = 250 − 8 = 242

9. D
$X^3 * X^2 = X^{3+2} = X^5$
To multiply exponents with like bases, add the exponents.

10. We first need to factorize 360 to its prime numbers:

$$\begin{array}{r|l} 360 & 2 \\ 180 & 2 \\ 90 & 2 \\ 45 & 3 \\ 15 & 3 \\ 5 & 5 \\ 1 & \end{array}$$

Then, $360 = 2^3 * 3^2 * 5^1$.

The number of positive factors of a number is found by the multiplication of the powers + 1 of the primes. Here, we need to multiply:

$(3 + 1)(2 + 1)(1 + 1) = 4 * 3 * 2 = 24$

Algebra

11. A
To simplify, remove the parenthesis and see if any terms cancel:

$(a + b)(x + y) + (a - b)(x - y) - (ax + by) = ax + ay + bx + by + ax - ay - bx + by - ax - by$

Writing similar terms together:

$= ax + ax - ax + bx - bx + ay - ay + by + by - by$ ... + terms cancel - terms:

$= ax + by$

12. C
We are asked to find A + B - C. By paying attention to the sign distribution, write the polynomials and operate:

$A + B - C = (-2x^4 + x^2 - 3x) + (x^4 - x^3 + 5) - (x^4 + 2x^3 + 4x + 5)$

$= -2x^4 + x^2 - 3x + x^4 - x^3 + 5 - x^4 - 2x^3 - 4x - 5$

$= -2x^4 + x^4 - x^4 - x^3 - 2x^3 + x^2 - 3x - 4x + 5 - 5$ ... similar terms written together for summing/substituting.

$= -2x^4 - 3x^3 + x^2 - 7x$

13. D
$x^2 + 4x + 4 = 0$ ... We try to separate the middle term $4x$ to find common factors with x^2 and 4 separately:

$x^2 + 2x + 2x + 4 = 0$ ... Here, x is a common factor for x^2 and $2x$, and 2 is a common factor for $2x$ and 4:

$x(x + 2) + 2(x + 2) = 0$ ... Here, we have x times $x + 2$ and 2 times $x + 2$ summed up. This means that we have $x + 2$ times $x + 2$:

$(x + 2)(x + 2) = 0$

$(x + 2)^2 = 0$ ... This is true if only if $x + 2$ is equal to zero.

$x + 2 = 0$

$x = -2$

14. B
To solve the equation, first we need to arrange it to appear in the form $ax^2 + bx + c = 0$ by removing the denominator:

$x - 31/x = 0$ ... First, we enlarge the equation by x:

$x * x - 31 * x/x = 0$

$x^2 - 31 = 0$

The quadratic formula to find the roots of a quadratic equation is:

$x_{1,2} = (-b \pm \sqrt{\Delta}) / 2a$ where $\Delta = b^2 - 4ac$ and is called the discriminant of the quadratic equation.

In our question, the equation is $x^2 - 31 = 0$. By remembering the form $ax^2 + bx + c = 0$:

$a = 1, b = 0, c = -31$

So, we can find the discriminant first, and then the roots of the equation:

$\Delta = b^2 - 4ac = 0^2 - 4 * 1 * (-31) = 124$

$x_{1,2} = (-b \pm \sqrt{\Delta}) / 2a = (\pm\sqrt{124}) / 2 = (\pm\sqrt{4 * 31}) / 2 = (\pm 2\sqrt{31}) / 2$ ... Simplifying by 2:

$x_{1,2} = \pm\sqrt{31}$ ... This means that the roots are $\sqrt{31}$ and $-\sqrt{31}$.

15. B
Finding the x-intercepts of a function means that we need to equate the function to zero and find the roots of the equation:

$(x - 5)^2 - 9 = 9$

$(x - 5)^2 = 9$

$\sqrt{(x - 5)^2} = \sqrt{9}$

$x - 5 = 3 \rightarrow x = 8$

$x - 5 = -3 \rightarrow x = 2$

16. C
First, we need to write two equations separately:

$4x - y = 5$ (I)

$x + 2y = 8$ (II) ... Here, we can use two ways to solve the system. One is substitution method, the other one is linear elimination method:

1. Substitution Method:

Equation (I) gives us that $y = 4x - 5$. We insert this value of y into equation (II):

$x + 2(4x - 5) = 8$

$x + 8x - 10 = 8$

$9x - 10 = 8$

$9x = 18$

$x = 2$

By knowing $x = 2$, we can find the value of y by inserting $x = 2$ into either of the equations. Choose equation (I):

$4(2) - y = 5$

$8 - y = 5$

$8 - 5 = y$

$y = 3$ → solution is (2, 3)

2. Linear Elimination Method:

2 * / 4x - y = 5 ... by multiplying equation (I) by 2, we see that -2y will form; and y terms

x + 2y = 8 ... will be eliminated when summed with +2y in equation (II):

2 * / 4x - y = 5

+   x + 2y = 8

    8x - 2y = 10

+ x + 2y = 8 ... Summing side-by-side:

8x + x - 2y + 2y = 10 + 8 ... -2y and +2y eliminate

each other:

9x = 18

x = 2

By knowing x = 2, we can find the value of y by inserting x = 2 into either of the equations. Choose equation (I):

4(2) - y = 5

8 - y = 5

8 - 5 = y

y = 3 → solution is (2, 3)

17. A

-x - 7 = -3x - 9

-x + 3x =-9 + 7

2x = -2

x = (-2):2

x = -1

18. C

$3(x + 2) - 2(1 - x) = 4x + 5$

$3x + 6 - 2 + 2x = 4x + 5$

$5x + 4 = 4x + 5$

$5x - 4x = 5 - 4$

$x = 1$

19. D
In this question, we do not need to try to find the value of x. Notice that the numbers containing x as power are of base 2 both in the given and asked expressions. Find the value of 2x first:

$2^{x-1} = 3$

$2^{-1} * 2^x = 3$
$2^x = 3 * 2$
$2^x = 6$

The value of 8^x is required, so,
$8^x = (2^3)^x = 2^{3x} = 6^3 = 216$

20. C
We can either try the equations in the answer choices, or without seeing them, we can directly calculate the equation of the graph. We observe that the graph intersects x-axis at (- 7, 0) and intersects y-axis at (0, 2). Then, trying these two points; we see that equation given in C is correct.

If we try to obtain the equation using the intersections; we need to use the formula:

$(y - y_1) / (y_1 - y_2) = (x - x_1) / (x_1 - x_2)$ where (x_1, y_1) = (-7, 0) and (x_2, y_2) = (0, 2)

$(y - 0) / (0 - 2) = (x - (-7)) / (-7 - 0)$

$y / (-2) = (x + 7) / (-7)$

$2(x + 7) = 7y$

$2x + 14 = 7y$

$2x - 7y + 14 = 0$

21. B

Noticing degree of 7 in the expression may make us think that this is a difficult problem. However, it is not. It is essential here to notice that the expressions of which 7th degree power is taken are negatives of each other. Remember that if n is odd and a is a positive number:

$(-a)^n = (-1)^n a^n = -a^n$.

So, $(y - x)^7 = (-1)^7 (x - y)^7 = -(x - y)^7$

$((x - y)^7 + x^2 + y^2 + (y - x)^7 + 2xy) / (y^2 - x^2)$

$= ((x - y)^7 + x^2 + y^2 - (x - y)^7 + 2xy) / (y^2 - x^2)$

$= (x^2 + y^2 + 2xy) / (y^2 - x^2)$

Now, we have a simplified form. Notice that the numerator is a perfect square; it is the square of $(x + y)$. Also, the numerator is the difference of two squares:

$= (x + y)^2 / ((y - x)(y + x))$

Simplify by $(x + y)$:

$= (x + y) / (y - x)$

22. D

The general equation of a circle is $(x - a)^2 + (y - b)^2 = r^2$ where (a, b) is the center of the circle and r is the radius. In the question, we are given that $a = 2$, $b = 5$ and $r = 3$. Inserting these values:

$(x - 2)^2 + (y - 5)^2 = 3^2$

$x^2 - 4x + 4 + y^2 - 10y + 25 = 9$

$x^2 + y^2 - 4x - 10y + 20 = 0$

23. D

More than one x value can result in the same y value, but a x value cannot result in different y values in a set of data or a graph to represent a function.
Check all answer choices:

a. {(0, 2), (1, 5), (5, 5), (3, 0)}
Notice that when x = 0, y = 2; when x = 1 and 5, y = 5 and when x = 3, y = 0. As mentioned above, two different x values can give the same y value but the reverse is not possible for a function. Therefore, x = 1 and x = 5 giving y = 5 is possible for a function; the other pairs already fit function properties.
b. $y = 4x^2$ is a function; we can find corresponding y values by inserting any x value. Since this is an even function; $f(-x) = f(x)$.

c. If we have a graph, the easiest way to check whether it represents a function or not is to draw vertical lines, parallel to y-axis. If the line drawn intersects with the graph at more than one point, then it is not a function:

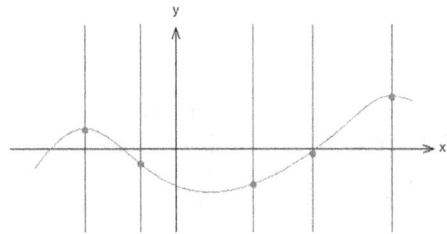

The red lines intersect with the graph at only one point, so this is the graph of a function.

d. Applying the same method, we see that vertical lines intersect with the graph at two points.

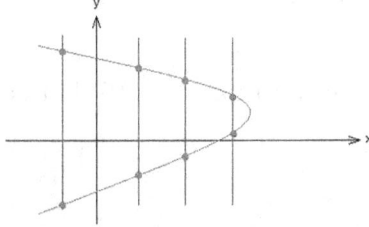

This means that the graph is not a function.

24. B

Noticing degree of 7 in the expression may make us think that this is a difficult problem. However, it is not. It is essential here to notice that the expressions of which the 7th degree power is taken are negatives of each other. Remember that if n is odd and a is a positive number, then,

$(-a)^n = (-1)^n a^n = -a^n$.

So, $(y - x)^7 = (-1)^7(x - y)^7 = -(x - y)^7$
$((x - y)^7 + x^2 + y^2 + (y - x)^7 + 2xy) / (y^2 - x^2)$

$= ((x - y)^7 + x^2 + y^2 - (x - y)^7 + 2xy) / (y^2 - x^2)$

$= (x^2 + y^2 + 2xy) / (y^2 - x^2)$

Now, we have a simplified form. Notice that the numerator is a perfect square; it is the square of $(x + y)$. Also, the numerator is the difference of two squares:

$= (x + y)^2 / ((y - x)(y + x))$
Simplify by $(x + y)$:

$= (x + y) / (y - x)$

25. D

In the table, x values are 2, 4, 6, 8 and y values are 6, 24, 54, 96. While x values increase by 2, the increase within sequent y values get higher by increasing x values. This means that there is a coefficient and/or exponential relationship between x and y.

Compare the four data to determine their dependence:

x: 2 → 4: multiplied by 2 ; y: 6 → 24: multiplied by 4 that is the square of 2

x: 2 → 6: multiplied by 3 ; y: 6 → 54: multiplied by 9 that is the square of 3

x: 2 → 8: multiplied by 4 ; y: 6 → 96: multiplied by 16 that is the square of 4

So, $y = ax^2$. Now, we should find a:

$x = 2 \rightarrow y = a * 2^2 = 6 \rightarrow a = 6/4 = 3/2$

Justify this by checking another point:

$x = 8 \rightarrow y = a * 8^2 = 96 \rightarrow a = 96/64 = 3/2$

Then; $y = (3/2)x^2$

26. C

If the function is $f(x) = x$; this will be definable for every x. However, there are some cases where we need to eliminate some ranges of x. In this question, there is a square root operation and a denominator.

The expression inside the square root cannot be negative. So:

$x + 7 \geq 0$

$x \geq -7$

On the other hand; the denominator cannot be zero since number divided by zero is not definable:

$x - 3 \neq 0$

$x \neq 3$

We have two limitations; x can neither be smaller than -7 nor equal to 3. So, the domain for this function is:

$[-7, +\infty) - \{3\}$ or we can also show by: $[-7, 3) \cup (3, +\infty)$.

27. C

First, determine the characteristics of the sequence. Note that the difference between 1st and 2nd terms is 3. However, the difference between 2nd and 3rd terms is 6. So, this is not an arithmetic sequence. The factor between 1st and 2nd term is 2. This factor is 2 for 2nd and 3rd terms as well. So, this is a geometric sequence. The sum formula for a geometric sequence is given by:

$S_n = a_1 (1 - r^n) / (1 - r)$ where S_n is the sum of the terms up to the nth term. We are asked to find the sum up to 10th term. r is the common ratio that is the factor between two successive terms; it is 2 for this question. a_1 is the first term of the sequence that is 3 according to the given data. So, inserting these values:

$S_n = 3 (1 - 2^{10}) / (1 - 2) = 3 (-1023) / (-1) = 3 * 1023 = 3069$

Note that the manual way to solve this problem is to write the whole sequence up to the requested term and sum.

28. D
The range of a graph is the set of y values of the function. Examining the graph above, we see that for all x values, y values are equal and above - 4. In other words; y cannot have values smaller than - 4. So, the range is [- 4, +∞).

29. A
Start by writing some values of the function:

$x = 1 : f(1) = 3$

$x = 2 : f(2) = 2 * f(1)$

$x = 3 : f(3) = 3 * f(2) = 3 * 2 * f(1)$

$x = 4 : f(4) = 4 * f(3) = 4 * 3 * 2 * f(1)$

$x = 5 : f(5) = 5 * f(4) = 5 * 4 * 3 * 2 * f(1)$

Notice that the expansion of f(n) contains n! times f(1). So,

$f(n) = n! * f(1)$

$f(n) = 3 * n!$

is the general formula of the function. Then,

$f(20) = 3 * 20!$

30. A
This is a logarithm question which requires the application of many identities. First, notice that we can apply $\log a^b = b \log a$ in the denominator:

$(\log_x y / \log_{xz} y^3) - \log_x {}^3\sqrt{z} = (\log_x y / 3\log_{xz} y) - \log_x {}^3\sqrt{z}$

Then, we need to apply base change $\log_a b = \log_c b / \log_c a$ in the denominator to make the numerator and the denominator similar, and prepare a form to have the possibility of simplification:

$= (\log_x y / (3\log_x y / \log_x xz)) - \log_x {}^3\sqrt{z}$

Now, by simplification:

$= (\log_x xz) / 3 - \log_x {}^3\sqrt{z}$

Now, organise the second term:

$= (\log_x xz) / 3 - \log_x z^{1/3}$

$= (\log_x xz) / 3 - \log_x z / 3$

Remember that $\log(a*b) = \log a + \log b$:

$= (\log_x x + \log_x z - \log_x z) / 3$

$= \log_x x / 3$

We know that $\log_a a = 1$:

$= 1/3$

31. D

Talking about periodic functions, we need to determine the period and the amplitude first. Then, we need to decide the type of the trigonometric function to be used. Here, we know that at time t = 0, the spring is stretched 6 cm; so the ball is 32 - 6 = 26 cm above the floor, and at t = 4 seconds, it is again 26 cm above the floor and is ready to go up again. Period is the duration between two cases when the movement is in the same direction passing through the same point. So, period T = 4 seconds.

Now, we need to decide which trigonometric function to use. We know that the graph of sine is in the middle at t = 0. However, in this question; we have 26 cm at t = 0; that is the y-intercept. That is why, we use cosine.

We know that the period of cost is 2π. The general form of cosine in harmonic motion is coswt where

$w = 2\pi/t$. Then, the cosine with period 4 is found by:

$w = 2\pi/4 = \pi/2$

$\cos((\pi/2)t)$

The amplitude is the magnitude difference from the center of oscillation to negative and positive peaks; that is 6 cm and the center passes through 32 cm:

$H(t) = 32 - 6 \cos((\pi/2)t)$

at $t = 9.5$ seconds;

$H(9.5) = 32 - 6 \cos((\pi/2) 9.5) = 32 - 6 \cos(4\pi + 15\pi/20)$

$= 32 - 6 \cos 3\pi/4 = 32 - 6 (- \cos\pi/4)$

$= 32 + 6\sqrt{2} / 2$

$= 36.24$ cm

32. B
$f(x) = 2x + 5$

$g(x) = 5x + 2.$

$g \circ f = g(f(x)) = g(2x + 5) = 5(2x + 5) + 2 = 10x + 25 + 2$
$= 10x + 27$

33. D
$f(x) = 1 + x^2$

$f \circ f = f(f(x)) = f(1 + x^2) = 1 + (1 + x^2)^2 = 1 + 1 + 2x^2 + x^4$
$= 2 + 2x^2 + x^4$

34. B
$f(x) = 1 - x$

$f^{-1}(1 - x) = x$

1 - x = t

x = 1 - t

$f^{-1}(t) = 1 - t$
$f^{-1}(x) = 1 - x$

$f^{-1}(1/2) = 1 - 1/2 = 1/2$

35. A
f(x) = 5x

g(x) = 7 - 2x

f(x) - g(x) = 5x - (7 - 2x) = 5x - 7 + 2x = 7x - 7
$(f(x) - g(x))^{-1}(f(x) - g(x)) = x$

$(f(x) - g(x))^{-1}(7x - 7) = x$

7x - 7 = t

7x = t + 7

x = (t + 7)/7

$(f(t) - g(t))^{-1}(t) = (t + 7)/7$

$(f(x) - g(x))^{-1}(x) = (x + 7)/7$

$(f(x) - g(x))^{-1}(0) = (0 + 7)/7 = 1$

Subtest II Geometry and Data

1. B
There are three squares forming a right triangle in the middle. Two of the squares have the areas 81 m² and 144 m². Denote their sides a and b respectively:

a² = 81 and b² = 144. The length, which is asked, is the hypotenuse; a and b are the opposite and adjacent sides of the right angle. Using the Pythagorean Theorem, we can find the value of the side:

Pythagorean Theorem:

(Hypotenuse)² = (Opposite Side)² + (Adjacent Side)²

$h^2 = a^2 + b^2$

$a^2 = 81$ and $b^2 = 144$ are given. So,

$h^2 = 81 + 144$

$h^2 = 225$

$h = 15$ m

2. A
Yes, the triangles are congruence - a case of SSA:

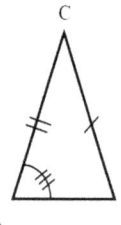 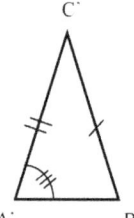

3.
We reflect the center O against the mirror line m at right angle and we use a compass to draw the circle with the same radius as the original circle.

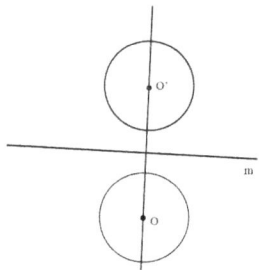

4. A
This is a case of ASA:

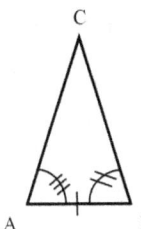 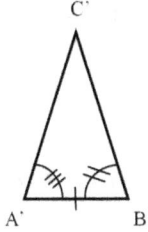

So the 2 triangles are congruent

5.

We reflect points C and D against the mirror line m at the right angle. Since points A and B are already on the mirror line, we can't reflect them and that's why A coincides with point A', and the same goes for points B and B'.

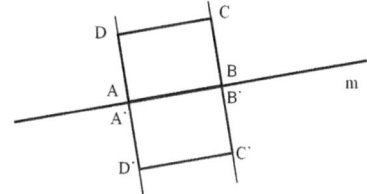

6. A

The formula for angle a is:

$a = (y - x) / 2 \rightarrow y - x = 2a \rightarrow y - x = 120$
and from the shape of the circle we also know that $x + y = 360$. Having two equations with two variables, we can find the value of x. To find it in the first step, we eliminate y:

$- / y - x = 120$

$y + x = 360$

$2x = 240$

$x = 120°$

7. C
The volume of a cone is 1/3 of the volume of a cylinder of the same height:

$$V = (\pi r^2 h) / 3$$

where r: radius, h: height. Insert the given values:

$$V = (\pi r^2 h) / 3 = (\pi * 5^2 * 12) / 3 = 100\pi \text{ cm}^3 = 100\pi * 10^{-6} \text{ m}^3$$

$$= \pi * 10^{-4} \text{m}^3$$

8. C

Visualizing, it is obvious that when the cones are cut, the following cross-section is created:

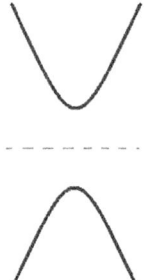

This is a hyperbola. The dashed line in the middle shows the location of the point where two vertexes unite.

When cut parallel to the floor; we obtain circles.

When cut at an angle (< 90º) with the floor, we obtain ellipses.

When only one cone is cut perpendicular or in an angle (< 90º) with the plane perpendicular to the floor, we obtain parabolas.

9. B
Total Volume = Volume of large cylinder - Volume of small cylinder

Volume of a cylinder = area of base X height = $\pi r^2 * h$

Total Volume = $(\pi * 12^2 * 10) - (\pi * 6^2 * 5) = 1440\pi - 180\pi$

$= 1260\pi \text{ in}^3$

Practice Test Questions 1

10. C

In the figure, we are given a large circle and a small circle inside, with the diameter equal to the radius of the large one. The diameter of the small circle is 4 cm. This means that its radius is 2 cm. Since the diameter of the small circle is the radius of the large circle, the radius of the large circle is 4 cm. The area of a circle is calculated by: $πr^2$ where r is the radius.

Area of the small circle: $π(2)^2 = 4π$

Area of the large circle: $π(4)^2 = 16π$

The difference area is found by:

Area of the large circle - Area of the small circle = $16π - 4π = 12π$

11. B

In the question, we have a right triangle formed inside the circle. We are asked to find the length of the hypotenuse of this triangle. We can find the other two sides of the triangle by using circle properties:

The diameter of the circle is equal to 12 cm. The legs of the right triangle are the radii of the circle; so they are 6 cm long.

Using the Pythagorean Theorem:

$(Hypotenuse)^2 = (Perpendicular)^2 + (Base)^2$
$h^2 = a^2 + b^2$

Given: d (diameter) = 12 & r (radius) = a = b = 6
$h^2 = a^2 + b^2$
$h^2 = 6^2 + 6^2$, $h^2 = 36 + 36$
$h^2 = 72$
$h = \sqrt{72}$
$h = 8.48$

12. A

The formula of the volume of cylinder is the base area multiplied by the height. As the formula:

Volume of a cylinder = $πr^2h$. Where π is 3.142, r is radius of the cross sectional area, and h is the height.

We know that the diameter is 5 meters, so the radius is 5/2 = 2.5 meters.

The volume is: V = 3.142 * 2.5² * 12 = 235.65 m³.

13. D
1 foot is equal to 12 inches. So 1 ft² = 12 * 12 in²

4 ft² = 4 * 12 * 12 in² = 576 in²

The surface area is divided equally among 3 boys.

Each boy will clean 576/3 = 192 in²

192 in² = 144 in² + 48 in²; 144 in² = 1 ft²

So, each boy will clean 1 ft² and 48 in²

14. D
As shown in the figure, two parallel lines intersect a third line with angle of 75°.

x = 75° (corresponding angles)

x + y = 180° (supplementary angles) ... inserting the value of x here:

y = 180° - 75°
y = 105°

15. D
We have a circle given with diameter 8 cm and a square located within the circle. We are asked to find the area of the circle for which we only need to know the length of the radius that is the half of the diameter.
Area of circle = πr² ... r = 8/2 = 4 cm

Area of circle = π * 4²

= 16π cm² ... As we notice, the inner square has no role in this question.

Probability and Statistics

16. A
There are 8 balls in the bag in total. It is important that two balls are taken out of the bag one by one. We can first take the blue then the white, or first white, then the blue. So, we will have two possibilities to be summed up. Since the balls are taken consecutively, we should be careful with the total number of balls for each case:

First blue, then white ball:

There are 3 blue balls; so, having a blue ball is 3/8 possible. Then, we have 7 balls left in the bag. The possibility to have a white ball is 1/7.

P = (3/8) * (1/7) = 3/56

First white, then blue ball:

There is only 1 white ball; so, having a white ball is 1/8 possible. Then, we have 7 balls left in the bag. The possibility to have a blue ball is 3/7.

P = (1/8) * (3/7) = 3/56

Overall probability is:

3/56 + 3/56 = 3/28

17. B
First add all the numbers 100 + 1050 + 320 + 600 + 150 = 2220. Then divide by 5 (the number of data provided) = 2220/5 = 444.

18. A
Probability that the 1st ball is red: 4/11

Probability the 2nd ball is green: 5/10

Combined probability is 4/11 * 5/10 = 20/110 = 2/11

19. D
First arrange the numbers in a numerical sequence - 1,2,3,4,5,6,7,8,9, 10. Then find the middle number or

numbers. The middle numbers are 5 and 6. The median = 5 + 6/2 = 11/2 = 5.5

20. B
The mode, or most occurring number in the series (7, 9, 10, 13, 11, 7, 9, 19, 12, 11, 9, 7, 9, 10, 11) is 9.

21. D
Assume that the first book chosen is red. Since we need to choose the second book in green or blue, there are 10 possible books to be chosen out of 15 - 1(that is the red book chosen first) = 14 books. There are equal number of books in each color, so the results will be the same if we think that blue or green book is the first book.

So, the probability will be 10/14 = 5/7.

22. B
A pie graph is a pie of which slices usually show percentage or proportional data on only 1 variable. We can show the parts of a whole by using pie graphs.

A scatter plot is used to show the change of 2 variables dependent on the same value. So, we can compare 2 variables just on one plot.

A bar chart contains columns; they are labeled to represent a categorical variable; so, it is used to represent 1 variable data.

Histograms are similar to bar charts, the only differences are the labelling - in histograms value labelling is done; and the columns are adjacent with no spacing.

23. C
Choices A, B and D show positive and negative correlations between the two parameters; however, they have no perceptible support. However, as mentioned in choice C; it is logical that the heater sales increase due to tougher weather conditions, because the weather gets colder and people get colder than before, so they need to get warmer than before which can result in an increase in heater sales.

24. B
80 out of 120 expect to eat out 5 days next month. This information gives the proportion of people expecting to eat out to total number of people. However, not all employees participated the survey; so we accept that the random sample represents all employees:
If 80 out of 120 expect to eat out next month, how many employees out of 450 expect to eat out next month?

450 * 80 / 120 = 300 employees

25. B
Ignoring the order means this is a combination problem, not permutation. The reader will choose 3 books out of 4. So,

C(4, 3) = 4! / (3! * (4 - 3)!) = 4! / (3! * 1!) = 4

There are 4 different ways.

Ignoring the order means this is a combination problem, not permutation. The reader will choose 3 books out of 4. So,

C(4, 3) = 4! / (3! * (4 - 3)!) = 4! / (3! * 1!) = 4

There are 4 different ways.

Subtest III Calculus

1. A
Remember the integral of x^a is found by:
x^{a+1} / (a + 1)

Constant coefficients inside the integral can be directly taken out. So,

$$\int_{-3}^{2} 2x - 3\,dx = 2(x - 3 + 1 / (-3 + 1))\Big|_{-3}^{2} = -x - 2\Big|_{-3}^{2}$$
= -(2-2 - (-3)-2) = -(1/4 - 1/9)

= -5/36

2. A
The fundamental theorem of calculus mentions that, with f continuous on [a, b]:

If $F(x) = \int_a^x f(t)dt \rightarrow F'(x) = f(x)$

Notice that the limits of the integral in the question need to be changed to make x the upper limit:

$G(x) = \int_x^{\pi/4} \cos^2 3x\, dx = -\int_{\pi/4}^x \cos^2 3x\, dx$

We see that $G'(x) = -f(x)$.

Since $f(x) = -\cos^2 3x$, $G'(x) = -\cos^2 3x$.

$G'(\pi/6) = -\cos^2(3 * \pi/6) = -\cos^2(\pi/2) = 0$

3. C
First, expand the expression:

$\int_{-1}^3 (4g(x) - 5f(x))dx = \int_{-1}^3 4g(x)dx - \int_{-1}^3 5f(x)dx$

Notice that the limits of the integrals change from -1 to 3. In the given integrals, they change from -1 to 3 for f(x), but from 3 to -1 for g(x). We can do the following shifting:

$\int_3^{-1} g(x)dx = 8 \rightarrow \int_{-1}^3 g(x)dx = -8$

So:

$\int_{-1}^3 4g(x)dx - \int_{-1}^3 5f(x)dx = 4\int_{-1}^3 g(x)dx - 5\int_{-1}^3 f(x)dx$

$= 4(-8) - 5(-15) = -32 + 75 = 43$

4. C
The first derivative of a function is the slope of the tangent

line to the function. If the slope of this line is positive, then the line shows an increasing linear correspondence which means that the function is increasing as well. On the contrary; if the first derivative is negative, the slope of the tangent line is negative; then, it is decreasing linearly. So, the function is decreasing.

The second derivative of a function tells us if the first derivative of the function is increasing or decreasing. If the second derivative is positive, then the first derivative is increasing, so the slope of the tangent line is increasing. Consequently; the graph of f is concave up shaped. On the other hand; if the second derivative is negative, then the first derivative is decreasing, so the slope of the tangent line is decreasing. Consequently; the graph of f is concave down shaped.

5. D

Recall that the Taylor Series for $1 / (1 - x)$ is:

$$1 / (1 - x) = 1 + x^2 + x^3 + x^4 + \dots$$

$$= \sum_{n=0}^{\infty} x^n$$

for $-1 < x < 1$. This interval is important. We need to find the x value that satisfy $1 / (1 - x) = 5/4$. If x is within $-1 < x < 1$, we can proceed:

$1 / (1 - x) = 5/4$

$5 - 5x = 4$

$5x = 1$

$x = 1/5$

x is found to be within the interval, so we can continue solving. Sum the first 5 terms of the series. This means that n starts by 0, ends by 4 because n = 0 gives the first term:

$1 / (1 - x) = 1 + x^2 + x^3 + x^4 + \dots$

∞

$$= \sum_{n=0}^{} x^n$$

$$= \sum_{n=0}^{4} (1/5)^n = (1/5)^0 + (1/5)^1 + (1/5)^2 + (1/5)^3 + (1/5)^4$$

$$= 1 + 1/5 + 1/25 + 1/125 + 1/625 = 1.2496$$

Rounding to the nearest hundredths:

$1.2496 = 1.25$ that is the exact value of $5/4 = 1.25$

6. A
Note that the flow rate of water is V'(t) that is the volume change by time. We are asked to find the height change by time: h'(t). Since the radius is constant, only volume and height depend on time.

Remembering the volume of a cylinder:

$V = \pi r^2 h$

$V(t) = \pi r^2 h(t)$

$(d/dt)\, V(t) = \pi r^2\, (d/dt)\, h(t)$

$V' = \pi r^2 h'$

Inserting the given values:

$8 = \pi * 2^2 * h'$

$h' = 2/\pi$

7. C
First, reorganize the differential equation to prepare for integration:

$dy/dx = x^3/y^2$ ... by cross multiplication:

$y^2 * dy = x^3 * dx$

$\int y^2 * dy = \int x^3 * dx$

$y^3/3 = x^4/4 + C$ ... It is important not to forget the constant

C for indefinite integration.

$y^3 = 3x^4/4 + 3C$

$y = \sqrt[3]{(3x^4/4 + 3C)}$

8. B
The derivative of a function as limit is found by:

$f'(x) = \lim\limits_{x \to 0} (f(x + \Delta x) - f(x)) / \Delta x$

Here, $f(x) = 3x^2 - 2x + 7$

$\to f(x + \Delta x) = 3(x + \Delta x)^2 - 2(x + \Delta x) + 7 = 3x^2 + 6x\Delta x + (\Delta x)^2 - 2x - 2\Delta x + 7$

$f'(x) = \lim\limits_{\Delta x \to 0} (f(x + \Delta x) - f(x)) / \Delta x$

$= \lim\limits_{\Delta x \to 0} (3x^2 + 6x\Delta x + (\Delta x)^2 - 2x - 2\Delta x + 7 - 3x^2 + 2x - 7) / \Delta x$

$= \lim\limits_{\Delta x \to 0} (6x\Delta x + (\Delta x)^2 - 2\Delta x) / \Delta x$

$= \lim\limits_{\Delta x \to 0} (6x + \Delta x - 2)$

$= 6x - 2$

9. A
The fundamental theorem of calculus mentions that, with f continuous on [a, b]:

If $F(x) = \int_a^x f(t)dt \to F'(x) = f(x)$

$F(x) = \int_4^x x^5 dx$, we are asked to find $F'(3)$

Since $f(x) = x^5$, $F'(x) = x^5$

$F'(3) = 3^5 = 243$

10. D
The local minimum and maximum of a function are found by the second derivative test. First, we need to take the first derivative of the function:
$f(x) = 6x^3 - 36x + 23$
$f'(x) = 18x^2 - 36$

Check if there are any x values to make f ' zero:

$18x^2 - 36 = 0 \rightarrow x^2 = 2 \rightarrow x = \sqrt{2}$ and $x = -\sqrt{2}$

Then: $f(\sqrt{2}) = 0$ and $f(-\sqrt{2}) = 0$

Next step is to calculate f "(x) and insert $\sqrt{2}$ and $-\sqrt{2}$ to check if these values make the second derivative positive or negative:

$f''(x) = 36x$

$f''(\sqrt{2}) = 36\sqrt{2} > 0$ ... Then, $(\sqrt{2}, 36\sqrt{2})$ is a local minimum

$f''(-\sqrt{2}) = -36\sqrt{2} < 0$ ... Then, $(-\sqrt{2}, -36\sqrt{2})$ is a local maximum

11. D
First, find the integral of the function within the interval given:

$$\int_{-3}^{3} 2x^2 dx = 2(x^3/3) \Big|_{-3}^{3} = (2/3)(27 - (-27)) = 36$$

Now, draw the rectangles to be used in Riemann sum.

The formula for this method is: $\sum_{i=0}^{n-1} f(x_i)\Delta x$.

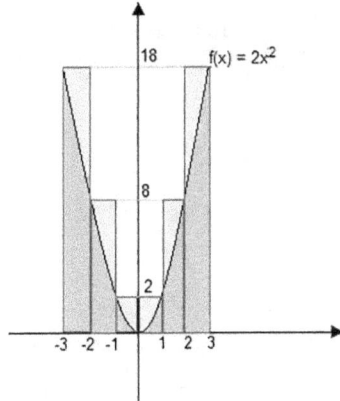

The smaller Δx value we use, the cleaner the calculation will be and the result will be closer to the integration result. In Riemann sum here, it is practical to find the total area between [0, 3] and multiply it by 2:

$$\sum_{i=0}^{n-1} f(x_i)\Delta x = 2[(1 - 0)*2(1)^2 + (2 - 1)*2(2)^2 + (3 - 2)*2(3)^2]$$

$$= 2[1*2 + 1*8 + 1*18] = 2*28 = 56$$

The difference of Riemann sum - integration is: 56 - 36 = 20

12. C
While dealing with limit problems, we first insert the limit of x into the function. If it contains indefinite cases, we carry on:

$$\lim_{x\to 5} (\tan(x^2 - 25) / (x - 5)) = \tan 0 / (5 - 5) = 0/0$$

Since this result is indefinite, we perform L'Hopital Rule. It says that in case we find $0/0$ or ∞/∞, we need to take the derivative of both numerator and denominator and then insert the limit of x:

$$\lim_{x\to 5} (\tan(x^2 - 25) / (x - 5)) = \lim_{x\to 5} (((d/dx) \tan(x^2 - 25)) / ((d/dx) (x - 5)))$$

$$= \lim_{x\to 5} (2x * \sec^2(x^2 - 25)) / 1 = 2 * 5 * \sec^2 0 = 10 * 1 = 10$$

13. A

The first order derivative of a function is equal to the slope of the tangent line. We are asked to find the equation of the formula. First, find the slope:

$y = 3x^3 - 8$

$y' = 9x^2$

at x = 1:

$y' = 9 * 1^2 = 9$ ... This is the slope.

We know that the general formula of a linear function is: y = mx + b.

We know that m = 9. We need to determine b.

Note that at x = 1, y = 3.1³ - 8 = -5. This means that at (1, -5), the line is tangent to the curve. So, this point is on the tangent line. Inserting the coordinates will give b:

-5 = 9 * 1 + b

b = -14

So the equation of the tangent line is: y = 9x - 14.

14. B

The values that cause the first derivative of the function to be zero or indefinite are the critical points. The local minimum and maximum values are at the critical points. However, not all the critical points are extremas.

Start by taking the first derivative:

$f(x) = x^3 - 27x$

$f' = 3x^2 - 27$

Notice that no x value makes f ' indefinite. So, next find the values that satisfy f ' = 0:

$f' = 3x^2 - 27 = 0$

$3x^2 = 27$

$x^2 = 9$

x = -3 and x = 3

So, -3 and 3 are the critical points. Now, we can get through 1st derivative test. Mark the critical points on a number line:

There are three regions here; (-∞, -3), (-3, 3) and (3, +∞). Next, select an integer from each interval and insert into the f ' equation to see whether it is negative of positive:

(-∞, -3): x = - 1: f ' = 3(-1)² - 27 = -24: negative

(-3, 3): x = 0: f ' = 3(0)² - 27 = -27: negative

(3, +∞): x = 4: f ' = 3(4)² - 27 = 21: positive

Now, insert the signs in the number line below:

This is the sign graph for the function. Notice that before and after -3; the graph is decreasing; there is no sign switch. So, there is no minimum or maximum at this point. However, the function decreases before 3, but increases after 3. This means that there is a local maximum at x = 3.

15. B

$$V_{Rotated\ Area} = \pi \int_0^{2\pi} (f(x))^2 dx$$

$$= \pi \int_0^{2\pi} (3 + \sin 2x)^2 dx$$

$$= \pi \int_0^{2\pi} (9 + 6\sin 2x + \sin^2 2x)\,dx$$

$$= 9\pi \int_0^{2\pi} dx + 6\pi \int_0^{2\pi} \sin 2x\,dx + \pi \int_0^{2\pi} \sin^2 2x\,dx$$

$$= 9\pi {}^* x \Big|_0^{2\pi} + 6\pi((-1/2)\cos 2x) \Big|_0^{2\pi} + \pi(x/2 - \sin 4x / 8) \Big|_0^{2\pi}$$

$$= 9\pi(2\pi - 0) - 3\pi(\cos 4\pi - \cos 0) + \pi((2\pi - 0)/2 - (\sin 8\pi - \sin 0)/8)$$

$$= 18\pi^2 - 0 + \pi(\pi - 0)$$

$$= 19\pi^2$$

This is the volume of the solid when rotated 2π degrees. If the graph is rotated $\pi/2$ degrees, it will scan a quarter of the total volume. That is:

$19\pi^2/4$

Practice Test Questions Set 2 (Difficult)

The questions below are not the same as you will find on the CSET® Mathematics test- that would be too easy! And nobody knows what the questions will be and they change all the time. Below are general questions that cover the same subject areas as the CSET® Mathematics test. So, while the format and exact wording of the questions may differ slightly, and change from year to year, if you can answer the questions below, you will have no problem with the CSET® Mathematics test.

For the best results, take these Practice Test Questions as if it were the real exam. Set aside time when you will not be disturbed, and a location that is quiet and free of distractions. Read the instructions carefully, read each question carefully, and answer to the best of your ability.
Use the bubble answer sheets provided. When you have completed the Practice Questions, check your answer against the Answer Key and read the explanation provided.

Do not attempt more than one set of practice test questions in one day. After completing the first practice test, wait two or three days before attempting the second set of questions.

Subtest I Answer Sheet

1. A B C D
2. A B C D
3. A B C D
4. A B C D
5. A B C D
6. A B C D
7. A B C D
8. A B C D
9. A B C D
10. A B C D
11. A B C D
12. A B C D
13. A B C D
14. A B C D
15. A B C D
16. A B C D
17. A B C D
18. A B C D
19. A B C D
20. A B C D
21. A B C D
22. A B C D
23. A B C D
24. A B C D
25. A B C D
26. A B C D
27. A B C D
28. A B C D
29. A B C D
30. A B C D
31. A B C D
32. A B C D
33. A B C D
34. A B C D
35. A B C D

Subtest II Answer Sheet

1. Ⓐ Ⓑ Ⓒ Ⓓ   18. Ⓐ Ⓑ Ⓒ Ⓓ
2. Ⓐ Ⓑ Ⓒ Ⓓ   19. Ⓐ Ⓑ Ⓒ Ⓓ
3. Ⓐ Ⓑ Ⓒ Ⓓ   20. Ⓐ Ⓑ Ⓒ Ⓓ
4. Ⓐ Ⓑ Ⓒ Ⓓ   21. Ⓐ Ⓑ Ⓒ Ⓓ
5. Ⓐ Ⓑ Ⓒ Ⓓ   22. Ⓐ Ⓑ Ⓒ Ⓓ
6. Ⓐ Ⓑ Ⓒ Ⓓ   23. Ⓐ Ⓑ Ⓒ Ⓓ
7. Ⓐ Ⓑ Ⓒ Ⓓ   24. Ⓐ Ⓑ Ⓒ Ⓓ
8. Ⓐ Ⓑ Ⓒ Ⓓ   25. Ⓐ Ⓑ Ⓒ Ⓓ
9. Ⓐ Ⓑ Ⓒ Ⓓ
10. Ⓐ Ⓑ Ⓒ Ⓓ
11. Ⓐ Ⓑ Ⓒ Ⓓ
12. Ⓐ Ⓑ Ⓒ Ⓓ
13. Ⓐ Ⓑ Ⓒ Ⓓ
14. Ⓐ Ⓑ Ⓒ Ⓓ
15. Ⓐ Ⓑ Ⓒ Ⓓ
16. Ⓐ Ⓑ Ⓒ Ⓓ
17. Ⓐ Ⓑ Ⓒ Ⓓ

Subtest III Answer Sheet

1. (A) (B) (C) (D)
2. (A) (B) (C) (D)
3. (A) (B) (C) (D)
4. (A) (B) (C) (D)
5. (A) (B) (C) (D)
6. (A) (B) (C) (D)
7. (A) (B) (C) (D)
8. (A) (B) (C) (D)
9. (A) (B) (C) (D)
10. (A) (B) (C) (D)
11. (A) (B) (C) (D)
12. (A) (B) (C) (D)
13. (A) (B) (C) (D)
14. (A) (B) (C) (D)
15. (A) (B) (C) (D)

Subtest I
Number and Quantity

1. 6/10 x 5/1

    a. 4/15
    b. 3/16
    c. 2 1/3
    d. 2/7

2. 2/15 ÷ 4/5

    a. 6/65
    b. 6/75
    c. 5/12
    d. 1/6

3. **Simplify $4^3 + 2^4$**

    a. 45
    b. 108
    c. 80
    d. 48

4. **If x = 2 and y = 5, solve $xy^3 - x^3$**

    a. 240
    b. 258
    c. 248
    d. 242

5. A gardener wants to plant trees around a rectangular garden with sides 32 m and 44 m. He wants to plant the trees with the largest equal gaps and one tree on each corner. How many seedlings does he need for this arrangement?

   a. 30
   b. 38
   c. 42
   d. 64

6. What is the remainder of 1! + 2! + 3! + ... + 100! when divided by 24?

   a. 7
   b. 9
   c. 11
   d. 13

7. $\sqrt[4]{2 * \sqrt[3]{4}} * \sqrt[3]{\sqrt{8}} = \sqrt[6]{4 * \sqrt[2]{2^{x+1}}}$ is given. Find the value of x.

   a. 2
   b. 3
   c. 5
   d. 6

8. Express 0.27 + 0.33 as a fraction.

   a. 3/6
   b. 4/7
   c. 3/5
   d. 2/7

9. Express 3^4 in standard form

　　a. 81
　　b. 27
　　c. 12
　　d. 9

10.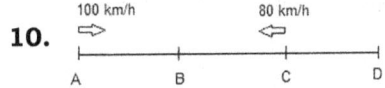

Two vehicles traveling at 100 km/h and 80 km/h start moving towards one another from cities A and C. When they start moving simultaneously, they meet in city B. If the vehicle with less velocity moved towards city D, two vehicles would meet in city D. Find the ratio |CD| / |BC|.

　　a. 5
　　b. 6
　　c. 8
　　d. 9

Algebra

11. Using the quadratic formula, solve the quadratic equation: $0.9x^2 + 1.8x - 2.7 = 0$

　　a. 1 and 3
　　b. -3 and 1
　　c. -3 and -1
　　d. -1 and 3

12. Factor the polynomial $x^2 - 7x - 30$.

   a. $(x + 15)(x - 2)$
   b. $(x + 10)(x - 3)$
   c. $(x - 10)(x + 3)$
   d. $(x - 15)(x + 2)$

13. $(3y^5 - 2y + y^4 + 2y^3 + 5) + (2y^5 + 3y^3 + 2 + 7y) =$

   a. $5y^5 + y^4 + 5y^3 + 5y + 7$
   b. $5y^3 + y^4 + 5y^3$
   c. $5y^5 + y^3 + 7y^3 + 5y + 5$
   d. $5y^2 + y^4 + 5y^3 + 7y + 5$

14. $(x^2 - 2)(3x^2 - 3x + 7) =$

   a. $3x^3 - 3x^3 + x^2 + 4x - 12$
   b. $3x^4 - 3x^3 + x^2 + 6x - 14$
   c. $3x^2 - 3x^3 + x + 6x - 10$
   d. $3x^2 - 3x + x + 4x - 14$

15. Find the solution for the following linear equation: $5x/2 = (3x + 24)/6$

   a. -1
   b. 0
   c. 1
   d. 2

16. If a and b are real numbers, solve the following equation: $(a + 2)x - b = -2 + (a + b)x$

   a. -1
   b. 0
   c. 1
   d. 2

17. Find the equation of the graph when $y = x^2$ is shifted 3 units left and 1 unit down.

    a. $y = x^2 + 2$
    b. $y = (x + 3)^2 + 1$
    c. $y = (x + 3)^2 - 1$
    d. $y = (x - 1)^2 - 3$

18. $A = 2^{2a-2}$ and $B = 5^{3-2a}$ are given. Find the value of 40^{4a-2} in terms of A and B.

    a. $2 * 10^4 * (A^4 / B^2)$
    b. $4 * 10^4 * (A^6 / B^2)$
    c. $4 * 10^8 * (A^6 / B^2)$
    d. $1* 6 * 10^5 * (A^6 / B^2)$

19. Find the area of the shape that is the limited by $|x + y + 1| \leq 3$ within the interval x: [-4, 0].

    a. 8 units²
    b. 12 units²
    c. 16 units²
    d. 24 units²

20. Given
$(1 - 1/2) * (1 - 1/3) * (1 - 1/4) \ldots . (1 - 1 / (x + 5)) = 1/18$, find the value of x.

    a. 5
    b. 8
    c. 13
    d. 18

21. In a microbiology laboratory, researchers are studying the growth properties of a newly discovered bacterium. It is known that the bacteria exhibits exponential growth. Initially, the researchers put $4 * 10^4$ bacteria into the growth medium. After 120 hours, they measure $1.024 * 10^{15}$ bacteria. Find the growth constant for the bacteria. Round your answer to the nearest tenths place.

   a. 0.1
   b. 0.2
   c. 0.3
   d. 0.4

22. Which of the following set of data does the graph below describe?

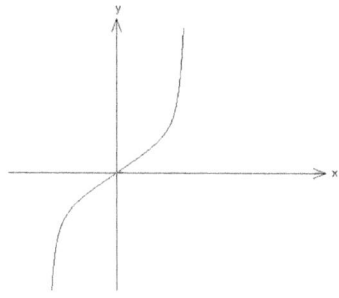

   a. f = {(-2, -8), (-1, -1), (0, 0), (1, 1), (2, 8)}
   b. f = {(-2, 8), (-1, 2), (0, 0), (1, 2), (2, 8)}
   c. f = {(-2, -11), (-1, -2), (0, 1), (1, -2), (2, -11)}
   d. f = {(-2, -1), (-1, 2), (0, 5), (1, 8), (2, 11)}

23. Find the equation of the graph below:

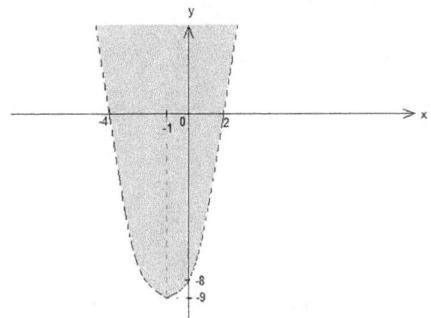

    a. $x^2 + 2x - 8 > 0$
    b. $2x + 8y \geq 0$
    c. $x^2 + 2x - 8 < 0$
    d. $2x - 8y < 0$

24. Solve the inequality $2(x + 1) - 3(-x - 1) < 2x - 1$.

    a. $x < 2$
    b. $x > 2$
    c. $x < -2$
    d. $x > 2$

25. Describe the end behavior for the function $f(x) = -2x^5 + 3x + 97$.

    a. Up on the left and right.
    b. Down on the left and right.
    c. Down on the left, up on the right.
    d. Up on the left, down on the right.

26. Determine the domain and the range of the table below:

x	y
-5	18
4	15
2	9
7	4
12	-3

a. domain: {- 3, 4, 9, 15, 18} range: {- 5, 4, 2, 7, 12}
b. domain: {9, 11, 11, 13, 19} range: {- 3, 4, 9, 15, 18}
c. domain: {- 5, 2, 4, 7, 12} range: {- 3, 4, 9, 15, 18}
d. domain: {9, 11, 13, 19} range: {- 3, 4, 9, 15, 18}

27. Find the domain of function given below:

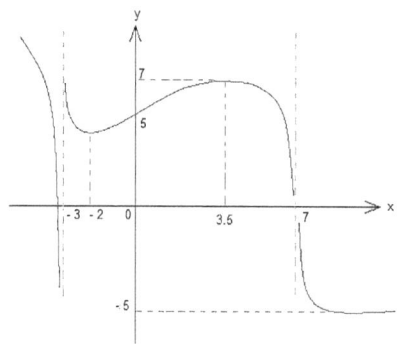

a. $(-\infty, +\infty) - \{-3, 7\}$
b. $(-\infty, +\infty) - \{-3, -2, 3.5, 7\}$
c. $(-\infty, +\infty) - \{-2, 3.5\}$
d. $(-\infty, +\infty) - \{-5, 5, 7\}$

28. Find the domain of the function $\sqrt{x-5}/(x+5)$.

   a. $(-\infty, 5]$
   b. $(-5, -\infty)$
   c. $[5, +\infty)$
   d. $[-\infty, +\infty)$

29. Find x and y from the following system of equations:

$(4x + 5y)/3 = ((x - 3y)/2) + 4$
$(3x + y)/2 = ((2x + 7y)/3) - 1$

   a. (1, 3)
   b. (2, 1)
   c. (1, 1)
   d. (0, 1)

30. The initial value for function f is given by f(1) = 3. The general formula of this function is f(x) = x * f(x - 1).

What is the value of f(20)?

   a. 3*20!
   b. 20^3
   c. 20*21
   d. 600

31. What is the result of $(\log_x y / \log_{xz} y^3) - \log_x \sqrt[3]{z}$?

   a. 1/3
   b. 1/27
   c. 3
   d. 9

32. A ball 32 cm above the floor is attached to the end of a spring attached to the ceiling. Initially, we pull the ball 6 cm down and when we let it move, it performs one up and down motion in 4 seconds. Modeling this harmonic movement using trigonometric functions (Assume that there is no air friction), find the distance between the ball and the ceiling at t = 9.5 seconds. Round your answer to the nearest hundredths.

   a. 28.12 cm

   b. 30.28 cm

   c. 32.36 cm

   d. 36.24 cm

33. Find $f^{-1}(1/2)$ if $f(x) = 1 - x$.

   a. 1

   b. 1/2

   c. 1/3

   d. 1/4

34. If $f(x) = 5x$ and $g(x) = 7 - 2x$, find $(f - g)^{-1}(0)$.

   a. 1

   b. 2

   c. 3

   d. 4

35. If $f^{-1}(x) = 2x$, find $f(x)$.

   a. x

   b. 2x

   c. x/2

   d. x/3

Subtest II
Geometry and Data

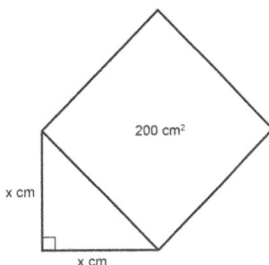

Note: Figure not drawn to scale

1. What is the length of the sides in the triangle above? Assume the quadrangle in the figure above is a square.

    a. 10
    b. 20
    c. 100
    d. 40

2. Given a right triangle where a is 12 and sinα=12/13, find cosα.

    a. -5/13
    b. -1/13
    c. 1/13
    d. 5/13

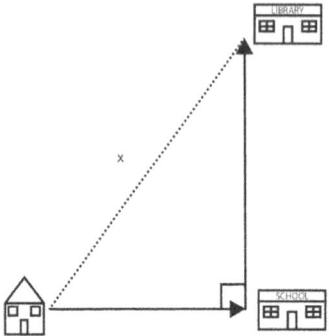

Note: Figure not drawn to scale

3. Every day starting from his home Peter travels due east 3 kilometers to the school. After school he travels due north 4 kilometers to the library. What is the distance between Peter's home and the library?

    a. 15 km
    b. 10 km
    c. 5 km
    d. 12 ½ km

4. Reflect the triangle ABC with the given mirror line m in the space below.

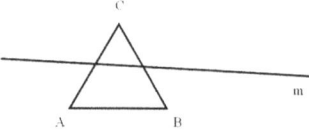

5. Reflect the quadrilateral ABCD in the coordinate plane if the mirror line is y-axis.

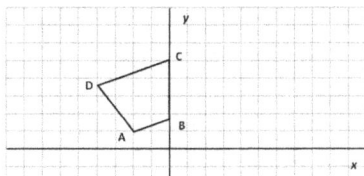

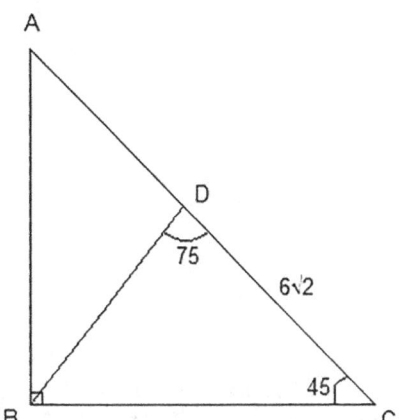

6. In the right isosceles triangles above: |CD| = 6√2 cm, m ∠ BCD = 45°, m ∠ CDB = 75°. Find the length of |AD|.

    a. 3√2 cm
    b. 2√6 cm
    c. 3√6 cm
    d. 6√2 cm

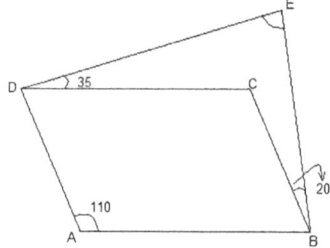

7. ABCD is a parallelogram. Find the value of m ∠ BED.

    a. 35°
    b. 45°
    c. 55°
    d. 65°

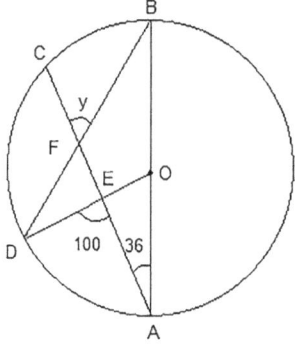

8. In the above figure, O is the center of the circle; B, O, A and A, E, F, C and D, E, O are linear, m ∠ DEA = 100°, m ∠ CAB = 36°. Find the value of m ∠ CFB = y.

    a. 58°
    b. 64°
    c. 68°
    d. 76°

9. The coordinates of the endpoints of a line are (a, b) and (3a, - 2b). If the midpoint of this line is (6, - 8), find a - b.

   a. -20
   b. -18
   c. -13
   d. 0

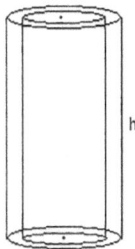

10. You are constructing a pipe by carving a solid cylinder with radius 5r, height 18r out a cylinder of radius 2r; its center overlapping with the outer cylinder. Find the total surface area of the pipe.

   a. 294 πr²
   b. 306 r²
   c. 325 πr²
   d. 340 πr²

11. Find the sum of all asymptotes of the function (2x² - 5x + 4) / (x² - 25).

   a. -7
   b. -5
   c. -3
   d. 2

12. Peter buys a conic hat, but it does not fit his head. He decides to cut the hat. He measures 16 cm from the vertex and marks it. Then moves 180° by the x-axis and when he reaches the opposite side of the previous mark, he marks 24 cm down the vertex. He joins the two marks with his pen and cuts along the circular line. After that, he tries on the hat. Find the equation of the conic section that Peter cut. Note that the vertex angle of the conic hat is 60°.

    a. $(13 - 4\sqrt{3})x^2 + 64y^2 = 1092 + 336\sqrt{3}$
    b. $(13 + 4\sqrt{3})x^2 + 84y^2 = 5824 + 1792\sqrt{3}$
    c. $4\sqrt{3}x^2 + 84y^2 = 2184 + 36\sqrt{3}$
    d. $(13 - 4\sqrt{3})x^2 - 84y^2 = 1092 + 1792\sqrt{3}$

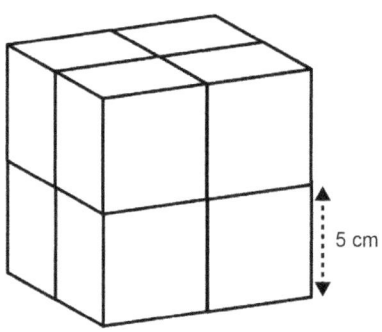

Note: Figure not drawn to scale

13. Assuming the figure above is composed of cubes, what is the volume?

    a. 125 cm³
    b. 875 cm³
    c. 1000 cm³
    d. 500 cm³

14. **Find the center of the hyperbola represented by the equation $64y^2 - 25x_2 - 384y - 100x - 1124 = 0$.**

    a. (-2, -3)
    b. (-2, 3)
    c. (3, -2)
    d. (3, 2)

15. **Rectangles ABCD and A'B'C'D' have following equal elements:**

Are these 2 rectangles congruent?

    a. Yes
    b. No
    c. Not enough information

Probability and Statistics

16. **There is a die and a coin. The dice is rolled and the coin is flipped according to the number the die is rolled. If the die is rolled only once, what is the probability of 4 successive heads?**

    a. 3/64
    b. 1/16
    c. 3/16
    d. 1/4

17. **These numbers are taken from the number of people that attended church every Friday for 7 weeks: 62, 18, 39, 13, 16, 37, 25. Find the mean.**

    a. 25
    b. 210
    c. 62
    d. 30

18. Smith and Simon are playing a card game. Smith will win if the drawn card form the deck of 52 is either 7 or a diamond, and Simon will win if the drawn card is an even number. Which statement is more likely to be correct?

    a. Smith will win more games.
    b. Simon will win more games.
    c. They have same winning probability.
    d. Decision could not be made from the provided data.

19. Find the mode from these test results: 90, 80, 77, 86, 90, 91, 77, 66, 69, 65, 43, 65, 75, 43, 90

    a. 43
    b. 77
    c. 65
    d. 90

20. A box contains 30 red, green and blue balls. The probability of drawing a red ball is twice the other colors due to its size. The number of green balls are 3 more than twice the number of blue balls, and blue are 5 less than the twice the red. What is the probability that 1st two balls drawn from the box randomly will be red?

    a. 10/102
    b. 11/102
    c. 1/29
    d. 1/30

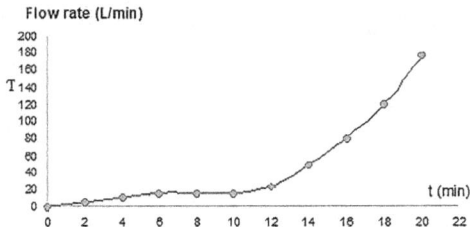

21. The scatter plot above shows the change of the water flow rate pouring from the tap by time. Which of the following statements is correct?

    a. The flow is linear between [0, 4] minutes.

    b. The tap is closed between [8, 10] minutes.

    c. The flow increases with higher slope between [12, 20] minutes than [0, 6] minutes.

    d. The flow rate linearly increases by time.

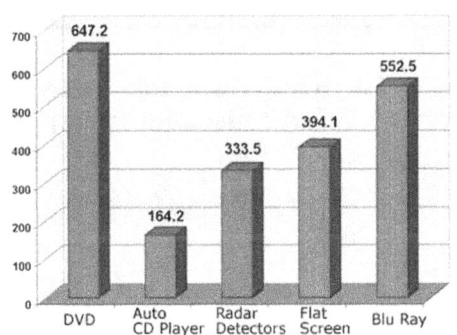

**22. Consider the graph above.
What is the third best-selling product?**

    a. Radar Detectors

    b. Flat Screen

    c. Blu Ray

    d. Auto CD Players

23. Given a normal distribution, what is the difference between the maximum value and the minimum value?

    a. Distribution
    b. Range
    c. Mode
    d. Median

24. In a museum, there are 250 visitors. An interviewer asks 50 people the number of days they visit museums per year. Here is the data obtained:

# of days museum visited per year	# of visitors interviewed
5	12
3	18
12	8
20	2
4	10

Based on the data, what is the most reasonable estimate for the number of visitors who visit museums 20 days in a year?

    a. 2
    b. 8
    c. 10
    d. 12

25. Sarah has two children and we know that she has a daughter. What is the probability that the other child is a girl as well?

    a. 1/4
    b. 1/3
    c. 1/2
    d. 1

Subtest III
Calculus

1. Find the difference between the integral of $f(x) = 2x^2$ and the area under this graph of this function using the Riemann sum with $\Delta x = 1$, within the interval $[-3, 3]$.

    a. 12
    b. 15
    c. 18
    d. 20

2. Given $F(x) = \int_{1}^{x}(2t + 1)dt$, which of the following is the equation of the line that is tangent to $F(x)$ at $x = 2$?

    a. $y = 2x + 5$
    b. $y = 5x - 6$
    c. $y = 6x + 5$
    d. $y = 7x - 2$

3. Using the definition of definite integral, compute $\int_{0}^{4} x^2 dx$.

    a. 64/3
    b. 32
    c. 48
    d. 64

4. Given the graph of f'(x); which of the following graphs represent the plot of function f(x)?

a.
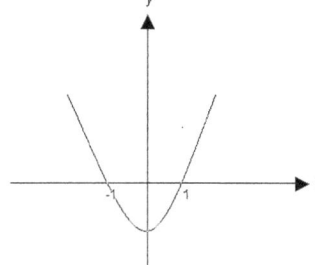

b.
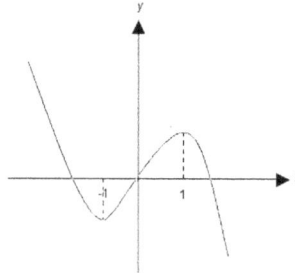

c.
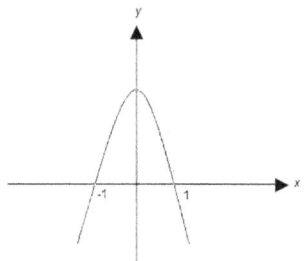

d.
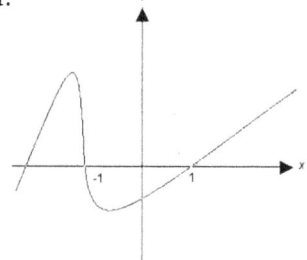

5. Find the interval of convergence of the power series

$$\sum_{n=1}^{\infty} ((x - 1)^n / (2n + 1)).$$

    a. $-1 < x < 1$
    b. $-1 < x < 2$
    c. $0 < x < 2$
    d. $1 < x < 3$

6. Jane and Kevin are standing 5 m apart. Jane starts walking north with a constant velocity and so, θ changes at a constant rate of 50/min. Find the rate of distance in m/min between Jane and Kevin when θ = 35°. Round your answer to the nearest hundredths.

   a. 15.40m/min
   b. 21.37 m/min
   c. 35.45 m/min
   d. 48.76 m/min

7. Given dy/dx = x^2y - 2y and y(0) = 1, find the value of the integration constant.

   a. -1
   b. 0
   c. 2
   d. 4

8. Using the limit definition, compute the derivative of f(x) = cos5x.

   a. 5cos5x
   b. -5cos5x
   c. 5sin5x
   d. -5sin5x

9. Given F(x) = $\int_{\pi/4}^{x}$ cos²t dt, which of the following is the equation of the line that is tangent to F(x) at x = π/2?

   a. y = (π + 1) - x/4
   b. y = y = (π + 1) / 4 + x
   c. y = (π - 1) / 4
   d. y = x/4 - π/8

10. Find the inflection point of the function $-(1/6)x^3 - x^2$.

    a. -4
    b. -2
    c. 0
    d. 2

11. Find the difference between the integral of $f(x) = 2x^2$ and the area under this graph of this function using the Riemann sum with $\Delta x = 1$, within the interval [-3, 3].

    a. 12
    b. 15
    c. 18
    d. 20

12. Find the limit of the function $(\sin 3x * (2x^2 + 3x - 5)) / (6x^2 + 15x)$ while $x \to 0$.

    a. -1
    b. 1
    c. 3
    d. 15

13. Find the equation of the line tangent to the curve $y = 3x^3 - 8$ at $x = 1$.

    a. $y = 9x - 14$
    b. $y = -9x + 14$
    c. $y = 9x + 16$
    d. $y = 18x + 16$

14. Find the local maximum of the function $f(x) = x^3 - 27x$.

    a. 1
    b. 3
    c. 18
    d. 27

15. If the graph of $f(x) = 3 + \sin^2 x$ is rotated $\pi/2$ degrees around the x-axis, what will the volume of the forming solid be within $[0, 2\pi]$?

    a. $7\pi^2/12$
    b. $19\pi^2/4$
    c. $9\pi^2/2$
    d. $19\pi^2$

Answer Key

Subtest I Number and Quantity

1. B
Since there are common numerators and denominators to cancel out, we cancel out 6/10 x 5/16 to get 6/2 x 1/16 = 3/2 x 1/8, and then we multiply numerators and denominators to get 3/16

2. D
To divide fractions, we multiply the first fraction with the inverse of the second fraction. Therefore we have 2/15 x 5/4, (cancel out) = 1/3 x ½ = 1/6

3. C
(4 x 4 x 4) + (2 x 2 x 2 x 2) = 64 + 16 = 80

4. D
$2(5)^3 - (2)^3$ = 2(125) – 8 = 250 – 8 = 242

5. B

First, find the length of the gap between sequent trees, found by the greatest common divisor (gcd) of 32 and 44:

32 44 | 2

16 22 | 2

8 11 | there are no more common divisors

gcd (32, 44) = 22 = 4

Next, find the number of trees to be planted on each side of the garden:

On sides of 32 m, the number of trees to be planted is: 32/4 + 1 = 9

On sides of 44 m, the number of trees to be planted is: 44/4 + 1 = 12

The total number of trees on four sides: 9 + 9 + 12 + 12 = 42

Remember to subtract 4 from 42, because two sides intersect on a corner on which there should be only one tree. So, For four corners, there will be 4 intersections, therefore, we need to subtract 4:

42 - 4 = 38 seedlings should be ordered.

6. B
Notice that if sum of series of a number is divided by a number, we can consider the division of each number in the series separately:

1! = 1, this gives 1 as remainder when divided by 24.

2! = 2, this gives 2 as remainder when divided by 24.

3! = 6, this gives 6 as remainder when divided by 24.

4! = 24, this gives 0 as remainder when divided by 24.

After 4!; each following term contains 24 as a factor, so they will give zero as remainder. Then, we only need to evaluate the remainder of 1! + 2! + 3! when divided by 24:

1! + 2! + 3! = 1 + 2 + 6 = 9

9 divided by 24 gives 9 as remainder.

7. D
In this type of question with one root within the other, we need to reduce the expression to one root with one degree that is found by multiplying all degrees of roots that follow.

Meanwhile; while taking a number inside a root, we need to take its power that is the degree of the root:

$\sqrt[4]{2 * \sqrt[3]{4}} * \sqrt[3]{\sqrt{8}} = \sqrt[6]{4 * \sqrt[2]{2^{x+1}}}$

$= \sqrt[4.3]{2^3 * 4} * \sqrt[3.2]{8} = \sqrt[6.2]{4^2 * 2^{x+1}}$

Notice that every term is a power of 2, so write all of them in base 2:

$= \sqrt[12]{2^3 * 2^2} * \sqrt[6]{2^3} = \sqrt[12]{2^4 * 2^{x+1}}$

$$= \sqrt[12]{2^5} * \sqrt[6]{2^3} = \sqrt[12]{2^{x+5}}$$

$$= 2^{5/12} * 2^{3/6} = 2^{(x+5)/12}$$
$$= 2^{(5/12 + 1/2)} = 2^{(x+5)/12}$$

Now that the bases are the same, we can equate the powers:

$5/12 + 1/2 = (x + 5) / 12$

$(5 + 6) / 12 = (x + 5) / 12$

$11 = x + 5$
$x = 6$

8. C
$0.27 + 0.33 = 0.6 = 6/10$, simplifying $3/5$

9. A
$3 \times 3 \times 3 \times 3 = 81$

10. D
We need to write equations with the two data provided:
If they move towards each other, they meet in city B. Since the time information is the same for both vehicles, we can equate distance/velocity ratios. Name $|AB| = a$, $|BC| = b$ and $|CD| = c$:

$|AB| / 100 = |BC| / 80$

$a / 100 = b / 80 \rightarrow a / 5 = b / 4 \rightarrow 4a = 5b$

If both vehicles move towards city D, they meet there. Again, time information is the same for both vehicles, so:

$|CD| / 80 = |AD| / 100$

$c / 80 = (a + b + c) / 100 \rightarrow c / 4 = (a + b + c) / 5 \rightarrow 5c = 4a + 4b + 4c$

$\rightarrow 4a + 4b = c$

We are asked to find $|CD| / |BC|$ that is c / b:
We have:

$4a = 5b$ ... (I)

$4a + 4b = c$ ... (II)

Inserting equation (I) into equation (II):

$5b + 4b = c$

$9b = c$

So, $c / b = 9b / b = 9$

Algebra

11. B
To solve the equation, we need the equation in the form $ax^2 + bx + c = 0$.
$0.9x^2 + 1.8x - 2.7 = 0$ is already in this form.

The quadratic formula to find the roots of a quadratic equation is:

$x_{1,2} = (-b \pm \sqrt{\Delta}) / 2a$ where $\Delta = b^2 - 4ac$ and is called the discriminant of the quadratic equation.

In our question, the equation is $0.9x^2 + 1.8x - 2.7 = 0$. To eliminate the decimals, multiply the equation by 10:

$9x^2 + 18x - 27 = 0$ ... This equation can be simplified by 9 since each term contains 9:

$x^2 + 2x - 3 = 0$

By remembering the form $ax^2 + bx + c = 0$:

$a = 1, b = 2, c = -3$

So, we can find the discriminant first, and then the roots of the equation:

$\Delta = b^2 - 4ac = (2)^2 - 4 * 1 * (-3) = 4 + 12 = 16$

$x_{1,2} = (-b \pm \sqrt{\Delta}) / 2a = (-2 \pm \sqrt{16}) / 2 = (-2 \pm 4) / 2$

This means that the roots are,

$x_1 = (-2 - 4)/2 = -3$ and $x_2 = (-2 + 4)/2 = 1$

12. C
$x^2 - 7x - 30 = 0$ ... We try to separate the middle term $-7x$ to find common factors with x^2 and -30 separately:

$x^2 - 10x + 3x - 30 = 0$ ... Here, we see that x is a common factor for x^2 and $-10x$, and 3 is a common factor for $3x$ and -30:

$x(x - 10) + 3(x - 10) = 0$ ... Here, we have x times $x - 10$ and 3 times $x - 10$ summed up. This means that we have $x + 3$ times $x - 10$:

$(x + 3)(x - 10) = 0$ or $(x - 10)(x + 3) = 0$

13. A
Write in standard form $(3y^5 + y^4 + 2y^3 - 2y + 5) + (2y^5 + 3y^3 + 7y + 2)$
Arrange in columns of like terms and then add

$\quad\quad 3y^5 + y^4 + 2y^3 - 2y + 5$
$\quad\quad 2y^5 + 3y^3 + 7y + 2$

$\quad\quad 5y^5 + y^4 + 5y^3 + 5y + 7$

14. B
$(x^2 - 2)(3x^2 - 3x + 7) = ?$

$= x^2(3x^2 - 3x + 7) - 2(3x^2 - 3x + 7)$

$= x^2(3x^2) + x^2(-3x) + x 2(7) - 2(3x^2) - 2(-3x) - 2(7)$ (6 terms)

$= 3x^4 - 3x^3 + 7x^2 - 6x^2 + 6x - 14$

$= 3x^4 - 3x^3 + (7 - 6)x^2 + 6x - 14$

$= 3x^4 - 3x^3 + x^2 + 6x - 14$

15. D
$5x/2 = (3x + 24)/6$

$3 * 5x/2 = (3x + 24)/6$

$15x/6 = (3x + 24)/6$

$15x = 3x + 24$

$15x - 3x = 24$

$12x = 24$

$x = 24/12 = 2$

16. A
$(a + 2)x - b = -2 + (a + b)x$

$ax + 2x - b = -2 + ax + bx$

$ax + 2x - ax - bx = -2 + b$
$2x - bx = -2 + b$

$(2 - b)x = -(2 - b)$

$x = -(2 - b):(2 - b)$

$x = -1$

17. C
Left shifting on horizontal axis is added to x, right shifting is subtracted from x. Shifting upwards is added on y, downwards shifting is subtracted from y.
So: shifting 3 units left: $y = (x + 3)^2$
shifting 1 unit down: $y = (x + 3)^2 - 1$

18. B
First, we need to be able to write 40 in terms of 2 and 5. By factorisation, we have:
$40 = 2^3 * 5$
So; $40^{4a-2} = (2^3 * 5)^{4a-2} = 2^{12a} * 2^{-6} * 5^{4a} * 5^{-2}$

Now, write 2^{12a} and 5^{4a} in terms of A and B, respectively:

$A = 2^{2a-2} \to 2^{2a} = 2^2 * A \to 2^{12a} = (2^2 * A)^6 \to 2^{12a} = 2^{12} * A^6$
$B = 5^{3-2a} \to 5^{-2a} = 5^{-3} * B \to 5^{2a} = 5^3 / B \to 5^{4a} = (5^3 / B)^2 \to 5^{4a} = 5^6 / B^2$

Then;

$40^{4a-2} = 2^{12a} * 2^{-6} * 5^{4a} * 5^{-2} = 2^{12} * A^6 * 2^{-6} * (5^6 / B^2) * 5^{-2}$

$= (A^6 / B^2) * 2^6 * 5^4$
$= (A^6 / B^2) * 2^4 * 2^2 * 5^4$
$= (A^6 / B^2) * 10^4 * 2^2$
$= 4 * 10^4 * (A^6 / B^2)$

19. D

First, draw the graph of the inequality given.

$|x + y + 1| \leq 3$

$-3 \leq x + y + 1 \leq 3$

$-4 \leq x + y \leq 2$

The region is within the lines $-4 \leq x + y$ and $x + y \leq 2$:

$x + y = -4 \rightarrow$ This line passes through (0, -4) and (-4, 0)

$x + y = 2 \rightarrow$ This line passes through (0, 2) and (2, 0)

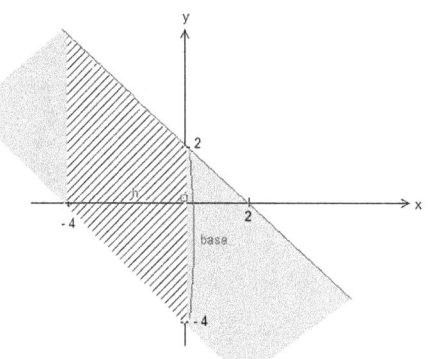

The grey + shaded area with limiting lines is the graph of $|x + y + 1| \leq 3$ and the shaded are on the grey area is the area within the interval x: [-4, 0].

This is a parallelogram with height of 4 units and (2 - (-4)) = 6 units base.

The area is: 4 * 6 = 24 units²

20. C

1 - 1/2 = 1/2
1 - 1/3 = 2/3
1 - 1/4 = 3/4

. .

. .

. .

1 - 1/(x + 5) = (x + 5 - 1) / (x + 5) = (x + 4) / (x + 5)

(1 - 1/2) * (1 - 1/3) * (1 - 1/4) * ... * (1 - 1 / (x + 5)) = 1/18

1/2 * 2/3 * 3/4 . ... . (x + 4) / (x + 5) = 1/18

Notice that starting from the first fraction; the denominator is cancelled by the numerator of the following fraction. After simplification, we have the following equation left:

1 / (x + 5) = 1/18

x + 5 = 18

x = 13

21. B

We are given that the bacteria reproduces exponentially; depending on time. So, we can write the growth function as:
$A = A_0 * e^{yt}$

where A is the number of bacteria at any time

A_0 is the initial number of bacteria; that is $4*10^4$ in this question

t is time in hours

y is the growth constant in hours[-1],

We know that at t = 120 hours, A = 1.024 * 10¹⁵ bacteria. Then,

$1.024 * 10^{15} = 4 * 10^4 * e^{y \cdot 120}$

$2^{10} * 10^{12} = 2^2 * 10^4 * e^{120y}$

$2^8 * 10^8 = e^{120y}$

$20^8 = e^{120y}$

$\ln 20^8 = \ln e^{120y}$

$8 * \ln 20 = 120y$

$y = \ln 20 / 15$

$y = 0.1997...$ when rounded to the nearest tenths: ≈ 0.2

22. A
The graph given belongs to an odd function because it is symmetric according to the origin. Then, we need to search for the set of data that represents an odd function.

Checking B and C, we see that f(-x) = f(x). Because the function values are the same for 2 and -2 and for 1 and -1. Then, these represent even functions.

Checking D, we notice that y values increase by 3 while x values increase by 1. This means that there is a linear relationship, so the function is a linear function. The graph given in the question is not linear.

So the answer is A. Notice that f(-x) = -f(x). Because the function values are the same in value, but opposite in sign for 2 and -2 and for 1 and -1. Then, this represents an odd function.

23. C
Analysing the graph.

Step 1: It shows quadratic characteristics, the roots of the function are x = -4 and x = 2. Answer choices B and D can be eliminated since they are linear functions.

Step 2: The graph is concave up which means that the term

x^2 has positive coefficient. The vertex passes through (-1, -9). The inequality does not cover 0 since the line is dashed. Point (0, 0) is not included in the inequality. By inserting x = 0 and y = 0, we can check whether A is the answer or C:

$x^2 + 2x - 8 > 0$

$0^2 + \cancel{2 * 0 - 8 > 0}$

-8 > 0 ... this is incorrect, so this is not the correct inequality.

$x^2 + 2x - 8 < 0$
$0^2 + 2 * 0 - 8 < 0$
-8 < 0 ... this is a correct statement. Yet, it is better that we check the other steps:

Step 1: Find the roots of $x^2 + 2x - 8 = 0$:

$x^2 + 2x - 8 = (x + 4)(x - 2) = 0$
x = -4 and x = 2. So, step 1 is proven.

Step 2:
$x^2 + 2x - 8 < 0$ ... the coefficient of term x^2 is 1, so the arms of the graph are concave up. The vertex (-1, -9) should satisfy $x^2 + 2x - 8 = 0$:

$(-1)^2 + 2(-1) - 8 = -9$

1 - 2 - 8 = -9

-9 = -9 ... The vertex is correct.

The inequality does not cover 0 since the line is dashed: $x^2 + 2x - 8 < 0$. Step 2 is proven.

24. C
2(x + 1) - 3(-x - 1) < 2x - 1
2x + 2 + 3x + 3 < 2x - 1
5x - 2x < -5 - 1
3x < -6
x < -2

25. D

Drawing the graph of a 5th degree polynomial is not a practical way to solve this problem. Instead, remember some properties of polynomial graphs:

Notice that this is an odd degree polynomial. So the two ends of the graph head off in opposite directions. If the leading term is positive; the left end would be down and the right end would be up.

However, the leading term here, $-2x^5$, is negative. So the end behavior for this function is up on the left, and down on the right.

26. C
The x values are the domain and the y values are the range. The domain of this set of data is {- 5, 2, 4, 7, 12} and the range is {- 3, 4, 9, 15, 18}.

27. D
The range of a graph is the set of y values of the function. Examining the graph above, we see that for all x values, y values are equal and above - 4. In other words; y cannot have values smaller than - 4. So, the range is [- 4, +∞).

28. D
The range of a graph is the set of y values of the function. Examining the graph above, we see that for all x values, y values are equal and above -4. In other words; y cannot have values smaller than -4. So, the range is [-4, +∞).

There are two points important in this question:

1. The term inside the square root cannot be smaller than zero which limits the range.

2. There are x values that make the denominator zero which make the function undefined. So, these values should be eliminated.

$\sqrt{(x - 5)} / (x + 5)$

$x - 5 \geq 0 \rightarrow x \geq 5$. Then, $x < 5$ should be eliminated.

$x + 5 \neq 0 \rightarrow x \neq -5$. Then, $x = -5$ should be eliminated.

Notice that -5 is already in $x < 5$. So, only excluding $x < 5$ values will be sufficient.

The domain of the function is: $[5, +\infty)$

29. C

First, we need to arrange the two equations to obtain the form $ax + by = c$. We see that there are 3 and 2 in the denominators of both equations. If we equate all at 6, then we can cancel all 6 in the denominators and have straight equations:

Equate all denominators at 6:

$2(4x + 5y)/6 = 3(x - 3y)/6 + 4 * 6/6$ ... Now we can cancel 6 in the denominators:

$8x + 10y = 3x - 9y + 24$ ... We can collect x and y terms on left side of the equation:

$8x + 10y - 3x + 9y = 24$

$5x + 19y = 24$ ... Equation (I)

Arrange the second equation:

$3(3x + y)/6 = 2(2x + 7y)/6 - 1 * 6/6$ ... Now we can cancel 6 in the denominators:

$9x + 3y = 4x + 14y - 6$ ... We can collect x and y terms on left side of the equation:

$9x + 3y - 4x - 14y = -6$

$5x - 11y = -6$ ... Equation (II)

Now, we have two equations and two unknowns x and y. By writing the two equations one under the other and operating, to find one unknown first, then the other:

$5x + 19y = 24$

-1/ 5x - 11y = -6 ... If we substitute this equation from the upper one, 5x cancels -5x:

5x + 19y = 24

-5x + 11y = 6 ... Summing side-by-side:

5x - 5x + 19y + 11y = 24 + 6

30y = 30 ... Dividing both sides by 30:

y = 1

Inserting y = 1 into either of the equations, we can find the value of x. Choosing equation I:

5x + 19 * 1 = 24

5x = 24 - 19

5x = 5 ... Dividing both sides by 5:

x = 1

So, x = 1 and y = 1 is the solution; it is shown as (1, 1).

30. A
Start by writing some values of the function:

x = 1 : f(1) = 3

x = 2 : f(2) = 2 * f(1)

x = 3 : f(3) = 3 * f(2) = 3 * 2 * f(1)

x = 4 : f(4) = 4 * f(3) = 4 * 3 * 2 * f(1)

x = 5 : f(5) = 5 * f(4) = 5 * 4 * 3 * 2 * f(1)

Notice that the expansion of f(n) contains n! times f(1). So,

f(n) = n! * f(1)

f(n) = 3 * n!

is the general formula of the function. Then,

$f(20) = 3 * 20!$

31. A
This is a logarithm question which requires the application of many identities. First, notice that we can apply $\log a^b = b \log a$ in the denominator:

$(\log_x y / \log_{xz} y^3) - \log_x{}^3\sqrt{z} = (\log_x y / 3\log_{xz} y) - \log_x{}^3\sqrt{z}$

Then, we need to apply base change $\log_a b = \log_c b / \log_c a$ in the denominator to make the numerator and the denominator similar, and prepare a form to have the possibility of simplification:

$= (\log_x y / (3\log_x y / \log_x xz)) - \log_x{}^3\sqrt{z}$

Now, by simplification:

$= (\log_x xz) / 3 - \log_x{}^3\sqrt{z}$

Now, organise the second term:

$= (\log_x xz) / 3 - \log_x z^{1/3}$

$= (\log_x xz) / 3 - \log_x z / 3$

Remember that $\log(a*b) = \log a + \log b$:

$= (\log_x x + \log_x z - \log_x z) / 3$

$= \log_x x / 3$

We know that $\log_a a = 1$:

$= 1/3$

32. D
When considering periodic functions, we need to determine the period and the amplitude first. Then, we need to decide the type of the trigonometric function to be used. Here, we know that at time $t = 0$, the spring is stretched 6 cm; so the ball is $32 - 6 = 26$ cm above the floor, and at $t = 4$ seconds, it is again 26 cm above the floor and is ready to go up again. Period is the duration between two cases when the movement is in the same direction passing through the same point. So, period $T = 4$ seconds.

Now, we need to decide which trigonometric function to use. We know that the graph of sine is in the middle at t = 0. However, in this question; we have 26 cm at t = 0; that is the y-intercept. That is why, we use cosine.

We know that the period of cost is 2π. The general form of cosine in harmonic motion is coswt where
w = $2\pi/t$. Then, the cosine with period 4 is found by:
w = $2\pi/4 = \pi/2$

cos(($\pi/2$)t)

The amplitude is the magnitude difference from the center of oscillation to negative and positive peaks; that is 6 cm and the center passes through 32 cm:

H(t) = 32 - 6 cos(($\pi/2$)t)

at t = 9.5 seconds;

H(9.5) = 32 - 6 cos(($\pi/2$) 9.5) = 32 - 6 cos ($4\pi + 15\pi/20$)

= 32 - 6 cos $3\pi/4$ = 32 - 6 (- cos$\pi/4$)

= 32 + 6$\sqrt{2}$ / 2

= 36.24 cm

33. B
f(x) = 1 - x

f^{-1}(1 - x) = x

1 - x = t

x = 1 - t

f^{-1}(t) = 1 - t

f^{-1}(x) = 1 - x

f^{-1}(1/2) = 1 - 1/2 = 1/2

34. A
f(x) = 5x

$g(x) = 7 - 2x$

$f(x) - g(x) = 5x - (7 - 2x) = 5x - 7 + 2x = 7x - 7$
$(f(x) - g(x))^{-1}(f(x) - g(x)) = x$

$(f(x) - g(x))^{-1}(7x - 7) = x$

$7x - 7 = t$

$7x = t + 7$

$x = (t + 7)/7$

$(f(t) - g(t))^{-1}(t) = (t + 7)/7$

$(f(x) - g(x))^{-1}(x) = (x + 7)/7$

$(f(x) - g(x))^{-1}(0) = (0 + 7)/7 = 1$

35. C
$f^{-1}(f(x)) = x$

Treat f(x) as a variable, so instead of 2x we have 2f(x).

$2f(x) = x$

$f(x) = x/2$

SubTest III Geometry and Data

1. A
If we call one side of the square "a," the area of the square will be a^2.

We know that $a^2 = 200$ cm^2.

On the other hand; there is an isosceles right triangle. Using the Pythagorean Theorem:

(Hypotenuse)^2 = (Adjacent Side)^2 + (Opposite Side)^2 Where the hypotenuse is equal to one side of the square. So,

$a^2 = x^2 + x^2$

$200 = 2x^2$

$200/2 = 2x^2/2$

$100 = x^2$

$x = \sqrt{100}$

$x = 10$ cm

2. D

To understand this question better, draw a right triangle by writing the given data on it:

Note: Figure not drawn to scale

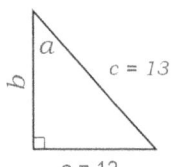

The side opposite angle a is named by a.

sin a = length of the opposite side / length of the hypotenuse = 12/13 is given.

cos a = length of the adjacent side / length of the hypotenuse = b/13

We use the Pythagorean Theorem to find the value of b:

(Hypotenuse)² = (Opposite Side)² + (Adjacent Side)²

$c^2 = a^2 + b^2$

$13^2 = 12^2 + b^2$

$169 = 144 + b^2$

$b^2 = 169 - 144$

$b^2 = 25$

$b = 5$

So,

cos a = b/13 = 5/13

3. C

We see that two legs of a right triangle form by Peter's movements and we are asked to find the length of the hypotenuse. We use the Pythagorean Theorem:

(Hypotenuse)² = (Adjacent side)² + (Opposite side)²

h² = a² + b²

We know that a and b are 3 km and 4 km. So,

h² = 3² + 4²

h² = 9 + 16

h² = 25

h = √25

h = 5 km

4.

We reflect points A, B and C against the mirror line m at the right angle and we connect the new points A', B' and C'. The process is the same even though the points of the triangle are not on the same side of the mirror line.

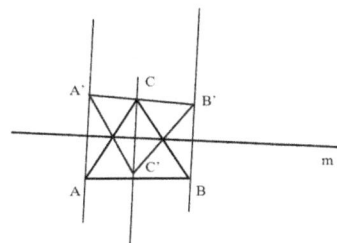

5.

The Reflect geometric shapes process is the same in the coordinate plane. Here, our mirror line is y-axis, so we reflect points A and D, and points B and C are already on the mirror line, so we don't reflect them.

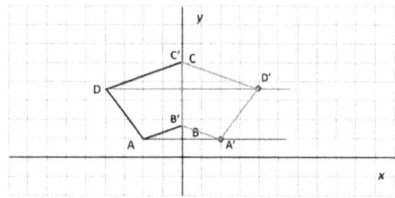

6. B

Drawing a height dividing 75°; we obtain a small triangle EDC that is similar to the large triangle BAC.

In a 45 - 45 - 90 triangle; the legs are $1/\sqrt{2}$ of the hypotenuse. So, |DE| = |EC| = 6 cm:

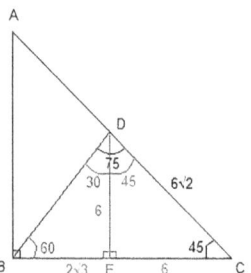

Then, in triangle BED; |BE| that is the opposite side of 300, is $1/\sqrt{3}$ of |ED| that is opposite to 600.

So, |BE| = $2\sqrt{3}$ cm.

ΔBAC ~ ΔEDC

Meaning that |EC| / |BE| = |DC| / |AD|

$6 / 2\sqrt{3} = 6\sqrt{2} / |AD|$ → |AD| = $2\sqrt{6}$ cm

7. C

In a parallelogram, opposite located angles have the same value.

So, m ∠ BCD = m ∠ DAB = 110°

m ∠ BCD = m ∠ EDC + m ∠ CBE + m ∠ BED is a shortcut geometric property for practical use:

110 = 35 + 20 + m ∠ BED

m ∠ BED = 110 - 55 = 55°

8. C

m ∠ CAB = 36° is a peripheral angle that makes the arc m(CB) = 36 . 2 = 72°.

Angles 100 and y are interior angles.

Angle y is found by the formula:

y = (m(CB) + m(DA)) / 2.
Inserting m(CB) = 72°, y = (72 + m(DA)) / 2. We need to find the measure of arc DA to find the measure of angle y. Use the other data given to find the missing value:

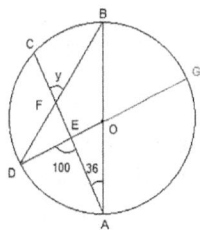

Similar to the previous operations, m ∠ DEA = 100° is found by the formula:

100 = (m(CG) + m(DA)) / 2
= (m(CB) + m(BG) + m(DA)) / 2
m(BG) = m(DA.

100 = (m(CB) + 2m(DA)) / 2

Inserting m(CB) = 72°:
100 = (72 + 2m(DA)) / 2

200 = 72 + 2m(DA)

m(DA. = (200 - 72) / 2 = 64°

Using this value, we can find y:
y = (72 + 64) / 2 = 136/2 = 68°

9. C
With the endpoints (x_1, y_1) and (x_2, y_2); the coordinates of the midpoint is found by:
$((x_1 + x_2) / 2, (y_1 + y_2) / 2)$

So, the midpoint of the line is:
((a + 3a. / 2, (b - 2b) / 2) = (2a, - b/2) = (6, - 8)

2a = 6 → a = 3

- b/2 = -8 → b = 16

Then; a - b = 3 - 16 = -13

10. B
0.45 kg = 1 pound, 1 kg. = 1/0.45 and 45 kg = 1/0.45 x 45 = 99.208, or 100 pounds.

The total surface area of a shape is found by summing all surfaces surrounding it. When a cylinder is carved to take a smaller cylinder out to obtain a pipe, we have three separate surfaces to calculate:

1. inner surface of the pipe
2. outer surface of the pipe
3. upper and lower areas of the pipe.

The inner and outer surfaces of the pipe are curled rectangles actually, and upper and lower areas are discs.

1. The inner surface of the pipe is the lateral area of the cylinder with radius 2r. The height of this rectangle is 18r as well; the width of this rectangle is the circumference of the circle with radius 2r:

Area of inner surface of the pipe: $(2\pi(2r)) * 18r = 72\pi r^2$

2. The outer surface of the pipe is the lateral area of the cylinder with radius 5r. The height of this rectangle is 18r, the width of this rectangle is the circumference of the circle with radius 5r:

Area of outer surface of the pipe: $(2\pi(5r)) \cdot 18r = 180\pi r^2$

3. The upper and lower areas of the pipe are discs with outer radius 5r and inner radius 2r:

The sum of upper and lower areas of the pipe:
$2|(\pi(5r)^2) - (\pi(2r)^2)| = 42\pi r^2$

The total surface area of the pipe:
$72\pi r^2 + 180\pi r^2 + 42\pi r^2 = 294\pi r^2$

11. D
The vertical asymptote is found by searching the values that make the function undefined. If there is any value to make the denominator zero, then it is the vertical asymptote:

$x^2 - 25 = 0$
$x = 5$ and $x = -5$

So $x = -5$ and $x = 5$ are the vertical asymptotes.

Notice that numerator and denominator are of the same degree (2). So, there is a non-zero horizontal asymptote and there is no slant asymptote.

Dividing the leading terms; we obtain the horizontal asymptote:

$2x^2 / x^2 = 2$

The sum of all asymptotes belonging to this function is $-5 + 5 + 2 = 2$.

12. B

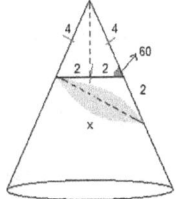

The circular that Peter marks make an ellipse. When cut among the marks, the hat will have the base as shown in the figure above (shaded). We need the dimensions of the ellipse to find the equation. The general equation of an ellipse is:
$x^2/a^2 + y^2/b^2 = 1$

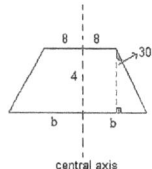

Note that all the circular and cross sections cut from a conic shape have centers on the same vertical axis. This information is important in finding b.

One end point of the circular shape 16 cm down the vertex overlaps with one end point of the elliptic shape cut. The other end point of the elliptic shape is 8 cm down the other end of the circular shape. It is obvious that the distance between the centers of the circular and elliptic shapes is 8/2 = 4 cm by simple ity. Then, find the value of b:

$b = 8 + 4/\sqrt{3} = (24 + 4\sqrt{3}) / 3$ cm

Now, we need to find 'a' that is the half of x in the figure. Using basic geometry information, we can find x.

Notice that the upper triangle is an equilateral triangle with side length 16 cm and interior angles 60°. Then; in the lower obtuse angle; we have one angle (180 - 60 = 120°) and two side of lengths 16 and 8 cm. There are many ways to find x. Apply the Law of Cosine to find x:

$x^2 = 2^2 + 4^2 + 2 * 2 * 4 * \cos 120$
$4 + 16 + 16 * (-\cos 60)$

=20 16/2
= 12 cm
a = x/2 = 6 cm

The equation of the ellipse is

$x^2/z^2 + y^2/b^2 = 1 \rightarrow x^2/6 + y^2/2 = 1/1$

$x^2 + 3y^2 = 6$

13. C
The large cube is made up of 8 smaller cubes with 5 cm sides. The volume of a cube is found by the third power of the length of one side.

Volume of the large cube = Volume of the small cube * 8

= $(5^3) * 8 = 125 * 8$

= 1000 cm^3

There is another solution for this question. Find the side length of the large cube. There are two cubes rows with 5 cm length for each. So, one side of the large cube is 10 cm.

The volume of this large cube is equal to 10^3 = 1000 cm^3

14. B

The general formula for a hyperbola is:

$(y - k)^2 / a^2 - (x - h)^2 / b^2 = 1$ where the center is (h, k).

Now, convert the equation in the question into the form above to obtain the coordinates of center:

$64y^2 - 25x^2 - 384y - 100x - 1124 = 0$

$64y^2 - 384y - 25x^2 - 100x - 1124 = 0$

$64(y^2 - 6y) - 25(x^2 + 4x) - 1124 = 0$

$64(y^2 - 6y + 9 - 9) - 25(x^2 + 4x + 4 - 4) - 1124 = 0$

$64(y - 3)^2 - 64 \cdot 9 - 25(x + 2)^2 + 25 * 4 - 1124 = 0$

$64(y - 3)^2 - 25(x + 2)^2 - 576 + 100 - 1124 = 0$

$64(y - 3)^2 - 25(x + 2)^2 = 1600$

$(64(y - 3)^2 - 25(x + 2)^2) / (64 \cdot 25) = 1600 / (64 \cdot 25)$

$(y - 3)^2 / 25 - (x + 2)^2 / 64 = 1$

$(y - 3)^2 / 5^2 - (x + 2)^2 / 8^2 = 1$

So; k = 3 and h = - 2. The center of the hyperbola is (- 2, 3).

15. A

We mark the elements that are the same:

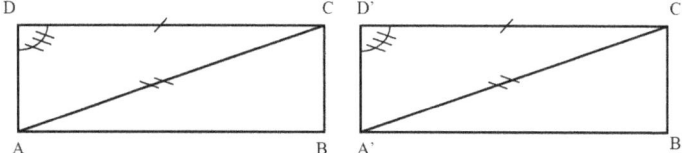

Angles at points D and D' are both right, because ABCD and A'B'C'D' are rectangles, so these 2 angles are equal. For triangles ACD and A'C'D' we have a case of SSA, so we have:

This means that sides AD and A'D' are equal. So, we can conclude that these 2 rectangles have all the same angles and same sides, so they are congruent.

Probability and Statistics

16. A

If the die is rolled for once, it can be 4, 5 or 6 since we are searching for 4 successive heads. We need to think each case separately. There are two possibilities for a coin; heads (H) or tails (T), each possibility of 1/2; we are searching for H. The possibility for a number to appear on the top of the die is 1/6. Die and coin cases are disjoint events. Also, each flip of coin is independent from the other:

Die: 4

coin: HHHH : 1 permutation

P = (1/6) * (1/2) * (1/2) * (1/2) * (1/2) = (1/6) * (1/16)

Die: 5

coin: HHHHT, THHHH, HHHHH : 3 permutations

P = (1/6) * 3 * (1/2) * (1/2) * (1/2) * (1/2) * (1/2) = (1/6) * (3/32)

Die: 6

coin: HHHHTT, TTHHHH, THHHHT, HHHHHT, THHHHH, HTHHHH, HHHHTH, HHHHHH : 8 permutations

P = (1/6) * 8 * (1/2) * (1/2) * (1/2) * (1/2) * (1/2) * (1/2) = (1/6) * (8/64)

The overall probability is:

Pall = (1/6) * (1/16) + (1/6) * (3/32) + (1/6) * (8/64)

= (1/6) * (1/16 + 3/32 + 8/64)

= (1/6) * (4 + 6 + 8) / 64 = (1/6) * (18/64) = 3/64

17. D
First add all the numbers 62 + 18 + 39 + 13 + 16 + 37 + 25 = 210. Then divide by 7 (the number of data provided) = 210/7 = 30

18. B
There are 52 cards in total. If we closely observe Smith has 16 cards in which he can win. So his winning probability in a single game will be 16/52 on the other hand Simon has 20 cards of wining so his probability on win in single draw is 20/52.

19. D
The most frequent occurring number in the series (90, 80, 77, 86, 90, 91, 77, 66, 69, 65, 43, 65, 75, 43, 90) is 90

20. A

Let the number of red balls be x

Then number of blue balls = 2x - 5

Then number of green balls= 2(2x - 5) + 3 = 4x - 10 + 3 = 4x - 7

As there are total 30 balls so the equation becomes

x + 2x - 5 + 4x - 7 = 30

x = 6

Red balls are 6, blue are 7 and green are 17.
As the probability of drawing a red ball is twice than the others, let's take them as 12. So the total number of balls will be 36.

Probability of drawing the 1st red: 12/36
Probability of drawing the 2nd red: 10/34

Combined probability = 12/36 X 10/34 = 10/102

21. C
Examining the plot; here is the information we get:
The flow rate increases linearly by time up to 6 minutes. Between 8 - 10 minutes, the flow rate is definitely constant around 20 L/min. After 12th minute; the flow starts increasing with a larger slope than it increases between 0 - 6 minutes. If the tap was closed at any time, we would observe a straight line on the horizontal axis.

22. B
Flat Screen TV are the third best-selling product.

23. B
The range of a distribution is the difference between the maximum value and the minimum value.

24. B
80 out of 120 expect to eat out 5 days next month. This information gives the proportion of people expecting to eat out to total number of people. However, not all employees par-

ticipated the survey; so we accept that the random sample represents all employees:

If 80 out of 120 expect to eat out next month, how many employees out of 450 expect to eat out?

450 * 80 / 120 = 300 employees

25. B

At first glance; we can think that a child can be either a girl or a boy, so the probability for the other child to be a girl is 1/2. However, we need to think deeper. The combinations of two children can be as follows:

boy + girl

boy + boy

girl + boy

girl + girl

So, the sample space is S = {BG, BB, GB, GG} where the sequence is important.

Sarah has a girl; this is the fact. So, calling this as event A, here are the possibilities:

boy + girl

girl + boy

girl + girl

We eliminate boy + boy, since one child is a girl. A = {BG, GB, GG}

The event that Sarah has two girls: B = {GG}

We need to compute:

$P(B|A) = P(B \cap A) / P(A) = 1/3$

Subtest III Calculus

1. D
First, find the integral of the function within the interval given:

$$\int_{-3}^{3} 2x^2 dx = 2(x^3/3)\Big|_{-3}^{3} = (2/3)(27 - (-27)) = 36$$

Now, draw the rectangles to be used in Riemann sum.

The formula for this method is: $\sum_{i=0}^{n-1} f(x_i)\Delta x$.

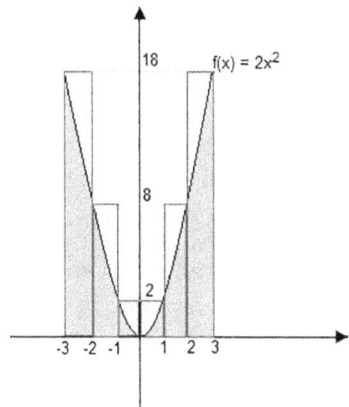

The smaller Δx value we use, the cleaner the calculation will be and the result will be closer to the integration result. In Riemann sum here, it is practical to find the total area between [0, 3] and multiply it by 2:

$$\sum_{i=0}^{n-1} f(x_i)\Delta x = 2[(1 - 0) * 2(1)^2 + (2 - 1) * 2(2)^2 + (3 - 2) * 2(3)^2]$$

$$= 2[1 * 2 + 1 * 8 + 1 * 18] = 2 * 28 = 56$$

The difference of Riemann sum - integration is: 56 - 36 = 20

2. B
The fundamental theorem of calculus mentions that, with f continuous on [a, b]:

1) If $F(x) = \int_a^x f(t)dt \rightarrow F'(x) = f(x)$

2) $\int_a^b f(t)dt = F(b) - F(a$. where, F is the antiderivative of f.

We need information at x = 2. Solve step-by-step:

1) If $F(2) = \int_1^2 (2t + 1)dt = (2t^2/2 + t) \Big|_1^2 = (2^2 - 1^2) + (2 - 1) = 4$

$F'(2) = f(2) \rightarrow F'(2) = 2 * 2 + 1 = 5$

The line that is tangent to F(x) at x = 2 passes through (2, 4) with slope 5. Find the equation of this line using the formula:

$y - y_1 = m(x - x_1)$

$y - 4 = 5(x - 2)$

$y = 5x - 6$

3. A
Determine the interval of integration first. Accepting that there are n slices to be sum:

$\Delta x = (b - a$. $/ n = (4 - 0) / 4 = 4/n$

The subintervals are: [0, 4/n], [4/n, 8/n], [8/n, 12/n], ..., [4(i - 1)/n, 4i/n], ..., [4(n - 1)/n, 4]

Notice that the right endpoint of the ith subinterval is:

$x_i^0 = 4i/n$

The integral is the sum of the function over the subinterval pieces:

$$\sum_{i=1}^{n} f(x_i^0)\Delta x = \sum_{i=1}^{n} [f(4i/n)] * (4/n) = \sum_{i=1}^{n} (4i/n)^2 * (4/n)$$

$$= \sum_{i=1}^{n} 64i^2 / n^3 = 64/n^3 \sum_{i=1}^{n} i^2$$

Remember the sum:

$$\sum_{i=1}^{n} i^2 = n(n + 1)(2n + 1) / 6$$

$$\sum_{i=1}^{n} f(x_i^0)\Delta x = (64/n^3) * n(n + 1)(2n + 1) / 6 = (64/6) *$$

$$((n + 1)(2n + 1)) / n^2$$

To find the value of definite integral, we need to take the limit of the above expression:

$(64/6) \lim_{n \to \infty} (((n + 1)(2n + 1)) / n^2) = (64/6) \lim_{n \to \infty} ((2n^2 + 3n + 1) / n^2) = (64/6) \lim_{n \to \infty} ((n^2(2 + 3/n + 1/n^2) / n^2))$

$= (64/6) * (2 + 0 + 0) = 64/3$

4. B

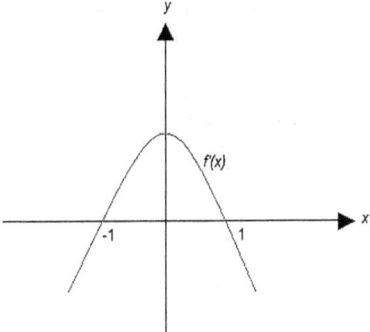

Checking the graph (above) of f'(x); we see that f' is negative when x < -1 and x > 1. This means that f is decreasing in this regions.

It is zero at x = -1 and x = 1 which mean that f has horizontal tangents at these points.

It is positive between (-1, 1). This means that f is increasing in this region.

Following these three notes, we see that the graph is as shown in choice B shown below.

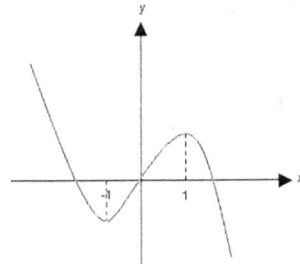

5. C

$\rho = \lim_{n \to \infty} |((x-1)^{n+1} / (2(n+1)+1)) / ((x-1)^n / (2n+1))|$

$= \lim_{n \to \infty} |(((x-1)^n * (x-1)) / (2n+3)) / ((x-1)^n / (2n+1))|$

$$= \lim_{n \to \infty} |((x-1)/(2n+3))/(1/(2n+1))|$$

$$= |x-1| * \lim_{n \to \infty} ((2n+1)/(2n+3))$$

$$= |x-1| * \lim_{n \to \infty} ((n(2+1/n))/(n(2+3/n))) = |x-1|$$

$-1 < x - 1 < 1$

$0 < x < 2$

The interval of convergence is $0 < x < 2$.

6. B

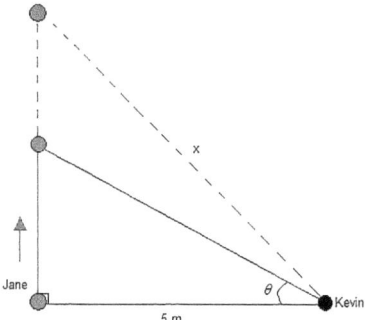

In the figure above, x is the distance between Jane and Kevin; we are asked to find the rate of change in x; that is x' when $\theta = 35°$.

Notice that both x and θ are time dependent: x(t) and θ(t).

Using right triangle properties:
$\cos \theta = 5/x \to x = 5/\cos \theta$

Now, derive both sides:

$(d/dt)x = d/dt \; 5/\cos \theta$

$x' = 5 \sin\theta * \theta/\cos^2 \theta$

We are given that the rate of change of the angle is $5°/min$: θ ' = $5°/min$ and $\theta = 35°$. Inserting these values:

x' = 5sin35 * 5/cos² 35 = 21.37m/min.

7. B
Noticing that the given differential equation is not ready to be integrated; we need to re-organise it by separation of variables:

$dy/dx = x^2y - 2y$

$dy/dx = y(x^2 - 2)$

$dy = y(x^2 - 2)dx$

$dy/y = (x^2 - 2)dx$ ... This form can be integrated:

$\int dy/y = \int (x^2 - 2)dx$

$\ln|y| = x^3/3 - 2x + C$

$|y| = e^{x^3/3 - 2x + C}$

Applying the initial value y(0) = 1 leads us to C:

$|y(0)| = e^{0 - 2.0 + C} = 1 \rightarrow e^C = 1 \rightarrow C = 0$

8. D
The derivative of a function as limit is found by:
$f'(x) = \lim_{\Delta x \to 0} (f(x + \Delta x) - f(x)) / \Delta x$

Here, $f(x) = \cos 5x$

$\rightarrow f(x + \Delta x) = \cos(5(x + \Delta x)) = \cos(5x + 5\Delta x)$

$f'(x) = \lim_{\Delta x \to 0} (f(x + \Delta x) - f(x)) / \Delta x = \lim_{\Delta x \to 0} (\cos(5x + 5\Delta x) - \cos 5x) / \Delta x$

Using the property: $\cos(a + b) = \cos a * \cos b - \sin a * \sin b$:

$= \lim_{\Delta x \to 0} (\cos 5x * \cos 5\Delta x - \sin 5x * \sin 5\Delta x - \cos 5x) / \Delta x$

$= \lim_{\Delta x \to 0} (\cos 5x(\cos 5\Delta x - 1) - \sin 5x * \sin 5\Delta x) / \Delta x$ ...

Recall that $\lim_{u \to 0} [(\cos u - 1) / u] = 0$ and $\lim_{u \to 0} (\sin u / u) = 1$

$= \lim_{\Delta x \to 0} [(\cos 5x (\cos 5\Delta x - 1)) / \Delta x] - \lim_{\Delta x \to 0} [(\sin 5x * \sin 5\Delta x) / \Delta x]$

$= \lim_{\Delta x \to 0} [(5\cos 5x (\cos 5\Delta x - 1)) / 5\Delta x] - \lim_{\Delta x \to 0} [(5\sin 5x * \sin 5\Delta x) / 5\Delta x]$

$= 5\cos 5x * \lim_{\Delta x \to 0} ((\cos 5\Delta x - 1) / 5\Delta x) - 5\sin 5x * \lim_{\Delta x \to 0} ((\sin 5\Delta x) / 5\Delta x)$

$= 5\cos x * 0 - 5\sin 5x * 1$

$= 0 - 5\sin 5x$

$= -5\sin 5x$

9. C
The fundamental theorem of calculus mentions that, with f continuous on [a, b]:

1) If $F(x) = \int_a^x f(t)dt \rightarrow F'(x) = f(x)$

2) $\int_a^b f(t)dt = F(b) - F(a)$ where, F is the antiderivative of f.

We need information at $x = \pi/2$. Solve step-by-step:

1) If $F(\pi/2) = \int_{\pi/4}^{\pi/2} \cos^2 t \, dt = (t/2 + \sin 2t / 4) \Big|_{\pi/4}^{\pi/2}$

$= (1/2) * ((\pi/2) - (\pi/4)) + (1/4) * (\sin(2 * \pi/2) - \sin(2 * \pi/4))$

$= \pi/4 - 1/4 = (\pi - 1) / 4$

$F'(\pi/2) = f(\pi/2) \rightarrow F'(\pi/2) = \cos^2(\pi/2) = 0$

The line that is tangent to F(x) at $x = \pi/2$ passes through $(\pi/2, (\pi - 1) / 4)$ with slope 0. Next, find the equation of this line using the formula:

$y - y_1 = m(x - x_1)$

$y - (\pi - 1) / 4 = 0(x - \pi/2)$

$y = (\pi - 1) / 4$

10. D
First, find the first derivative of the function:
$f(x) = - (1/6)x^3 - x^2$

$f'(x) = - (1/2)x^2 - 2x$ ... Now, find the x values where f' is zero:

$f'(x) = x((- 1/2)x - 2) = 0$

x = 0 and x = - 4

x:		-4		0	
sign of f ':	−	|	+	|	−
behaviour of the graph:	↓	|	↑	|	↓

Now, find the second derivative:
$f''(x) = -x - 2$ ... Now, find the x values where f'' is zero:

$f''(x) = -x - 2 = 0$

x = -2

x:		- 2	
sign of f '':	+	|	−
concavity of the graph:	up	|	down

Combine both charts to determine the graph behavior:

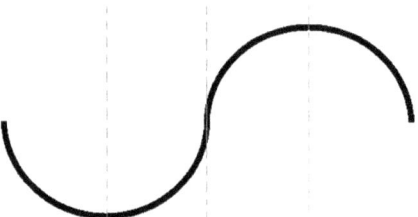

Notice that at x = 2, the graph changes its concavity; Meaning that, this is the inflection point.

11. D

First, find the integral of the function within the interval given:

$$\int_{-3}^{3} 2x^2 dx = 2(x^3/3) \Big|_{-3}^{3} = (2/3)(27 - (-27)) = 36$$

Now, draw the rectangles to be used in Riemann sum.

The formula for this method is: $\sum_{i=0}^{n-1} f(x_i)\Delta x$.

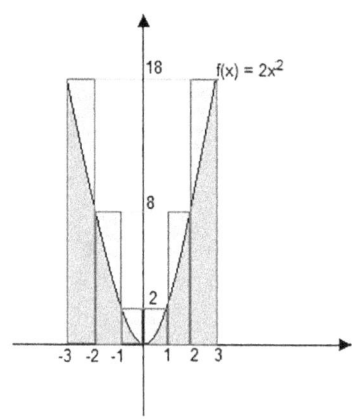

The smaller Δx value we use, the cleaner the calculation will be and the result will be closer to the integration result.

Practice Test Questions 2

In Riemann sum here, it is practical to find the total area between [0, 3] and multiply it by 2:

$$\sum_{i=0}^{n-1} f(x_i)\Delta x = 2[(1 - 0)*2(1)^2 + (2 - 1)*2(2)^2 + (3 - 2)*2(3)^2]$$

$$= 2[1 * 2 + 1 * 8 + 1 * 18] = 2 * 28 = 56$$

The difference of Riemann sum - integration is: 56 - 36 = 20

12. C
While dealing with limit problems, we first insert the limit of x into the function. If it contains indefinite cases, we carry on:

$$\lim_{x \to 5} (\tan(x^2 - 25) / (x - 5)) = \tan 0 / (5 - 5) = 0/0$$

Since this result is indefinite, we perform L'Hopital Rule. It says that in case we find $0/0$ or ∞/∞, we need to take the derivative of both numerator and denominator and then insert the limit of x:

$$\lim_{x \to 5} (\tan(x^2 - 25) / (x - 5)) = \lim_{x \to 5} (((d/dx) \tan(x^2 - 25)) / ((d/dx)(x - 5)))$$

$$= \lim_{x \to 5} (2x * \sec^2(x^2 - 25)) / 1 = 2 * 5 * \sec^2 0 = 10 * 1 = 10$$

13. A
The first order derivative of a function is equal to the slope of the tangent line. We are asked to find the equation of the formula. Start with finding the slope first:
$y = 3x^3 - 8$

$y' = 9x^2$

at x = 1:

$y' = 9 * 1^2 = 9$ ... This is the slope.

We know that the general formula of a linear function is:

$y = mx + b$.

We know that $m = 9$. We need to determine b.

Note that at $x = 1$, $y = 3 * 1^3 - 8 = -5$. This means that at (1, -5), the line is tangent to the curve. So, this point is on the tangent line. Inserting the coordinates will give b:

$-5 = 9 * 1 + b$

$b = -14$

So the equation of the tangent line is: $y = 9x - 14$.

14. B

The values that cause the first derivative of the function to be zero or indefinite are the critical points. The local minimum and maximum values are at the critical points. However, not all the critical points are extremas.

Start by taking the first derivative:

$f(x) = x^3 - 27x$

$f' = 3x^2 - 27$

Notice that no x value makes f ' indefinite. So, next find the values that satisfy f ' = 0:

$f' = 3x^2 - 27 = 0$

$3x^2 = 27$

$x^2 = 9$

$x = -3$ and $x = 3$

So, -3 and 3 are the critical points. Now, we can get through 1st derivative test. Let us mark the critical points on a number line:

There are three regions here; $(-\infty, -3)$, $(-3, 3)$ and $(3, +\infty)$.

Next, select an integer from each interval and insert into the f ' equation to see whether it is negative of positive:

$(-\infty, -3)$: $x = -1$: $f' = 3(-1)^2 - 27 = -24$: negative

$(-3, 3)$: $x = 0$: $f' = 3(0)^2 - 27 = -27$: negative

$(3, +\infty)$: $x = 4$: $f' = 3(4)^2 - 27 = 21$: positive

Now, insert the signs in the number line below:

This is the sign graph for the function. Notice that before and after -3; the graph is decreasing; there is no sign switch. So, there is no minimum or maximum at this point. However, the function decreases before 3, but increases after 3. This means that there is a local maximum at $x = 3$.

15. B

$$V_{Rotated\ Area} = \pi \int_0^{2\pi} (f(x))^2 dx$$

$$= \pi \int_0^{2\pi} (3 + \sin 2x)^2 dx$$

$$= \pi \int_0^{2\pi} (9 + 6\sin 2x + \sin^2 2x) dx$$

$$= 9\pi \int_0^{2\pi} dx + 6\pi \int_0^{2\pi} \sin 2x dx + \pi \int_0^{2\pi} \sin^2 2x dx$$

$$= 9\pi \, {}^*x \Big|_0^{2\pi} + 6\pi((-1/2)\cos 2x)\Big|_0^{2\pi} + \pi(x/2 - \sin 4x/8)\Big|_0^{2\pi}$$

$$= 9\pi(2\pi - 0) - 3\pi(\cos 4\pi - \cos 0) + \pi((2\pi - 0)/2 - (\sin 8\pi - \sin 0)/8)$$

$$= 18\pi^2 - 0 + \pi(\pi - 0)$$

$$= 19\pi^2$$

This is the volume of the solid when rotated 2π degrees. If the graph is rotated $\pi/2$ degrees, it will scan a quarter of the total volume. That is:

$19\pi^2/4$

Conclusion

CONGRATULATIONS! You have made it this far because you have applied yourself diligently to\potential score considerably! Getting into a good school is a huge step in a journey that might be challenging at times but will be many times more rewarding and fulfilling. That is why being prepared is so important.

www.ingramcontent.com/pod-product-compliance
Lightning Source LLC
Chambersburg PA
CBHW071852070526
44583CB00016B/1647